John Gould

An introduction to the Trochilidae or family of hummingbirds

John Gould

An introduction to the Trochilidae or family of hummingbirds

ISBN/EAN: 9783741171055

Manufactured in Europe, USA, Canada, Australia, Japa

Cover: Foto ©Thomas Meinert / pixelio.de

Manufactured and distributed by brebook publishing software (www.brebook.com)

John Gould

An introduction to the Trochilidae or family of hummingbirds

AN

INTRODUCTION

TO

THE TROCHILIDÆ,

OR

FAMILY OF HUMMING-BIRDS.

BY

JOHN GOULD, F.R.S., &c. &c.

LONDON:
PRINTED FOR THE AUTHOR,
BY TAYLOR AND FRANCIS, RED LION COURT, FLEET STREET.
1861.

[*The Author reserves to himself the right of Translation.*]

TO

HER ROYAL HIGHNESS

THE CROWN PRINCESS OF PRUSSIA,

PRINCESS ROYAL OF ENGLAND,

THIS WORK,

ON

THE TROCHILIDÆ,

OR

FAMILY OF HUMMING-BIRDS,

IS, WITH PERMISSION,

DEDICATED

BY HER ROYAL HIGHNESS'S

MOST OBEDIENT AND FAITHFUL SERVANT,

JOHN GOULD.

NOTICE.

As the Introduction to my "Monograph of the Trochilidæ" involved much intricate and laborious investigation, particularly with regard to the synonymy of the various species, I have been induced to have it set up in octavo for the facility of correction. From this draft, as it were, it has been reprinted in large type for the folio work. Believing that in its present form it might be interesting and useful to many of my scientific friends and others, I have had a limited number of copies printed for distribution among them. It must not, however, be regarded as a complete history of the family, but merely as an introduction to, and a revision of, the genera: the history of the species must be sought for in the folio work. At the same time it contains a considerable amount of information which has been acquired since the commencement of the publication, together with many additions to the synonymy; these are indicated by prefixed asterisks, the synonyms not so distinguished being merely an abbreviated reprint of those which have already appeared in the folio edition. As it is not to be expected that persons unconnected with science should be conversant with the abbreviations of the names of the authors and the titles of the

NOTICE.

works referred to, a fully detailed list of these has been added for their information.

In an early page I have stated that the family consisted of nearly 400 species; but it will be found that 416 are enumerated, 360 of which are figured. About 400 species are contained in my own collection, and these will be at all times accessible to men of science for the purposes of examination and comparison.

London, 20 Charlotte Street,
Bedford Square, W.C.,
Sept. 1, 1861.

PREFACE.

That early impressions of the mind are vividly retained, while events of the day flit from our memory, must have been experienced by every one. How vivid, then, is my recollection of the first Humming-Bird which met my admiring gaze! with what delight did I examine its tiny body and feast my eyes on its glittering plumage! This early impression, I well remember, gradually increased into an earnest desire to attain a more intimate acquaintance with the lovely group of birds to which it pertained, and was still further strengthened when an opportunity was afforded me of inspecting the, at that time, unique collection of the *Trochilidæ* formed by the late Mr. George Loddiges, of Hackney. This gentleman and myself were imbued with a kindred spirit in the love we both entertained for this family of living gems. To describe the feeling which animated us with regard to them is impossible; it can, in fact, only be realized by those who have made Natural History a study, and who pursue the investigation of its charming mysteries with ardour and delight. That our enthusiasm and excitement with regard to most things become lessened, if not deadened, by time, particularly when we have acquired what we vainly consider a complete knowledge of the subject, is, I fear, too often the case with most of us; not so, however, I believe with those who take up the study of the Family of Humming-Birds. Certainly I can affirm that such is not the case with myself; for the pleasure which I experience on seeing a Humming-Bird is as great at the present moment as when I first saw one. During the first twenty years of my acquaintance with these wonderful works of creation, my thoughts were often directed to them in the day, and my night dreams have not unfrequently carried me to their native forests in the distant country of America.

In passing through this world I have remarked that when inquirers of a strong will really set themselves to attain a definite object, they generally accomplish it; and in my own case the time at length arrived when I was permitted to revel in the delight of seeing the Humming-Birds in a state of nature, and to observe their habits in the woods and among the great flowering trees of the United States of America and in Canada. For some time a single Humming-Bird was my constant companion during days of toil by road and rail, and I ultimately succeeded in bringing a living pair within the confines of the British Islands, and a single individual to London, where it lived for two days, when, from the want of proper food or the change of climate, it died.

Although so enthusiastically attached to the subject, I should not have formed a collection of the *Trochilidæ*, or attempted an account of their history, had not my late friend Mr. George Loddiges (whose many excellences are too universally known to need any comment from me) been prematurely removed from among us. Prior to his lamented death, whatever species I procured from my various correspondents were freely placed at his disposal; and his collection was then unrivalled, and the pride of the owner as well as of his country, so far as a private collection could be considered of national importance. It was not until after Mr. Loddiges' decease that I determined upon forming the collection I myself possess, which now far surpasses every other, both in the number of species and examples. Ten years ago this collection was exhibited for a short time in the Gardens of the Zoological Society in the Regent's Park, and, I believe, afforded unmixed delight to the many thousands who visited those Gardens in the memorable year 1851. Many favourable notices of it appeared in the periodicals of the day; and my friend Mr. Martin published a small popular work in express reference to it. During the period which has since elapsed I have been unceasing in my endeavours to obtain every species which has been discovered by the enterprising travellers of this country, of Germany, of France, and of America. It would be invidious were I to extol the exertions of one more than those of another, nor could I do so without committing injustice; for the travellers of all these countries have shown equal intrepidity in their endeavours to bring to light the hidden treasures of the great primæval forests of the New World. Some of them, such as Azara, Spix, Bullock, De Lattre, Floresi, Dyson, Hoffmann, and Matthews (the discoverer of the wonderful *Loddigesia mirabilis*), are no longer among us: of those living who have paid especial attention to the Humming-Birds I may mention the names of Prince Maximilian of Wied, Waterton, Gosse, Warszewicz, Linden, Bridges, Jameson, Wallace, Bates, Darwin, Reeves, Hauxwell, Skinner, Bourcier, Sallé, Salvin, Fraser, Gundlach, Bryant, Montes de Oca, &c. It is to these men, living and dead, that science is indebted for a knowledge of so many of these "gems of creation;" and it is by their exertions that such collections as Mr. Loddiges' and my own have been formed. I regret exceedingly that I have not seen so much of this lovely group of birds in a state of nature as I could have wished: the traveller and the historian seldom go together; and in this instance it would have been impossible. The constant personal attention and care necessary for the production of such a work as 'A Monograph of the Trochilidæ' could only be given in a metropolis; for in no other place could such a publication be accomplished without a greatly increased expenditure both of time and money; it is only in capitals like London and Paris, that undertakings of this nature can be carried out successfully; for nowhere else are the requisite talents and materials to be obtained.

I feel that I am greatly indebted to those who have honoured this work with their support for their kindness and the patience with which they have continued with me to its completion—the more especially as, owing to the discovery of so many new species since its commencement, it has extended far beyond its expected limits. I am also especially indebted to those persons connected with its production, by whose assistance I have been enabled to bring so great an undertaking to a satisfactory close. To my artist Mr. Richter, to Mr. Prince, and to Mr. Bayfield (all names connected with my former works), I owe many thanks. To the projectors and publisher of 'Curtis's Botanical Magazine' I am indebted also for many hints and for permission to copy parts of some of their plates of the flowering plants of those districts of South America which are frequented by Humming-Birds. In case the merits of this work should be unknown to some of my subscribers, it is generally acknowledged that its production reflects equal credit upon its Editors Sir William Jackson Hooker and Mr. Smith, to the artist Mr. Fitch, and to its publisher Mr. Lovell Reeve.

Numerous attempts had been made at various times to give something like a representation of the glittering hues with which this group of birds are adorned, but all had ended in disappointment; and the subject seemed so fraught with difficulty that I at first despaired of its accomplishment. I determined, however, to make the trial, and, after a series of lengthened, troublesome, and costly experiments, I have, I trust, partially, if not completely succeeded. Similar attempts were simultaneously carried on in America by W. M. L. Baily, Esq., who with the utmost kindness and liberality explained his process to me; and although I have not adopted it, I must in fairness admit that it is fully as successful as my own. I shall always entertain a lively remembrance of the pleasant day I spent with this gentleman in Philadelphia. It was in his company that I first saw a living Humming-Bird in a garden which has become classic ground to all true Americans, from the pleasing associations connected with its former possessor, the great and good Bartram, and from its having been one of the haunts of the celebrated Wilson, than whom no one has written more pleasingly on the species of this family which inhabits that part of North America, the *Trochilus colubris*.

It now becomes my pleasing duty to place on record the very valuable assistance in the production of this work with which I have been favoured by the Directors of Public Museums and private individuals. Of these the foremost on the list must be the names of M. Jules Bourcier, of Paris, and Thomas Reeves, Esq., of Rio de Janeiro. Both these gentlemen have made extensive collections of specimens, and have had numerous drawings prepared for the express purpose of publishing works on the subject, which with the utmost liberality have been placed at my disposal. To M. Bourcier, than whom no one possesses a more intimate acquaintance with this group of birds, I am likewise indebted for much valuable information which has been at all times rendered with the utmost willingness and promptitude. My thanks are also due to the Trustees and the Keepers of the Zoological Department of the British Museum; to the Director of the Museum of the Jardin des Plantes at Paris; to Dr. Peters, Director of the Royal Zoological Museum of Berlin; to George Ure Skinner, Esq., long resident in Guatemala; to that intrepid traveller M. Warszewicz, now Director of the Botanic Garden at Cracow, who, during his travels in South America, brought to light more new species of Humming-Birds than any other explorer; to my friends Sir William Jardine, Bart.; W. C. L. Martin, Esq.; T. C. Eyton, Esq.; Dr. Sclater; Alfred Newton, Esq.; M. Edouard Verreaux, of Paris; G. N. Lawrence, Esq., of New York; and Dr. Baird, of Washington; to Edward Wilson, Esq., to Sigismund Rucker, Esq., F. Taylor, Esq., of Liverpool; William Tucker, Esq., of Trinidad; and to T. F. Erskine, Esq., for the readiness with which they have at all times favoured me with both information and the loan of specimens. To Miss Loddiges and her brother Mr. Conrad Loddiges, I am under considerable obligations for the facility of access they have always afforded me to the very valuable collection formed by their lamented father. Nor must the name of another valuable friend —the late Prince Charles Lucien Bonaparte—be omitted from the list of those who took great interest in the present work, he having at all times rendered me that scientific assistance which his vast and varied talents so well enabled him to afford.

September 1, 1861.

INTRODUCTION.

THE question has often been asked, whence the term Humming-Bird has been derived, why the bird is so called. I may state in reply that, owing to the rapid movement of the wings of most of the members of this group, but especially of the smaller species, a vibratory or humming sound is produced while the bird is in the air, which may be heard at the distance of several yards, and that it is from this circumstance that the trivial name by which these birds are known in England has arisen. In France they are recognised by the terms *Oiseau-Mouche* and *Colibri*; in Germany their common appellation is *Kolibri*; by the Dutch they are called *Kolibrielje*; by the Spaniards *Pica flores* and *Tomino*; by the Portuguese *Tomeneco* and *Beija-flor*; in the neighbourhood of Xalapa they are known by the names of *Chupa-rosa* and *Chupa-myrta*, Rose-sucker and Myrtle-sucker; by the Creoles of the Antilles and Guiana they are known by the names of *Murmures*, *Bourdons*, and *Frou-frous*. From the Mexicans, Peruvians, and other nations of South America they have received various appellations, such as *Ourissia*, *Auitzitzil*, *tzitztototl*, *guanumbi*, *quinti* or *quintiul*, *quindé*, *visicilin*, *pigda*, and *courbiri*; all terms of a metaphorical character, signifying "rays of the sun," "tresses of the day-star," "murmuring birds," &c.

Linnæus applied to the whole of the species known to him the generic appellation of *Trochilus*, a name given to some fabulous little bird by the ancients, and whence is derived the family designation of TROCHILIDÆ. By Brisson, a contemporary of Linnæus, the terms *Polytmus* and *Mellisuga* were proposed; but with respect to some of the thirty-six species described by him, as well as by the older writers, such as Seba, Marcgrave, Willoughby, Ray, &c., it is extremely difficult, if not impossible, to determine what they really were. We may, however, fairly commence our investigations with a greater chance of accuracy from the date when the great Swedish naturalist commenced his labours. By him twenty-two species were enumerated in the twelfth edition of his 'Systema Naturæ.' In Gmelin's, or the thirteenth edition, the list is increased to sixty-seven. Of these I have determined about two-thirds; the remainder must for ever continue involved in mystery, and their names be erased from our scientific works—the descriptions being extremely meagre, and the synonyms occasionally referring to figures of very different species. In some instances, even, the species are attributed to countries where Humming-Birds are never found; while in others, such as that of the Harlequin Humming-Bird, the characters are taken from a plate which must have been drawn from imagination and not from any real specimen. These are a few of the difficulties which a naturalist has to encounter when access to the types cannot be obtained. I think it necessary to make this statement as a reason for not quoting all the names given by the older authors. Wherever they could be with certainty determined, they have been quoted under the species to which they are believed to refer. The numerous divisions which more modern writers have deemed it necessary to propose will be given in the proper place.

Latham, who added little or nothing to the previously recorded notices of this group of birds, enumerated sixty-five species in his 'Index Ornithologicus,' published in 1790, and ninety-five in the third volume of his 'General History of Birds,' which appeared in 1822. Of these about two-thirds are real species, the remainder cannot be determined, as they are so indefinitely described that it is impossible to ascertain whether they are species or not.

In 1802 the 'Oiseaux dorés,' the great French work of Audebert and Vieillot, was given to the world. In it, besides figures of all the Jacamars and Promerops then known, were included seventy plates of Humming-Birds. These plates represent species which, though then rare, are now extremely common, and which, although not so numerous as those contained in the later work of Latham, had the advantage of being illustrated in a manner which was intended to convey some idea of their brilliancy. In most instances the species may be recognised; in others they are doubtful. Independently of the illustrations above-mentioned these authors attempted to explain the laws which produce the splendid colouring of certain parts of these beautiful birds, and have given a plate illustrative of their views on the subject.

In 1823 appeared the second part of the ornithological portion of the 'Tableau Encyclopédique et Méthodique des Trois Règnes de la Nature,' by Bonnaterre and Vieillot, with an enumeration of ninety-four species of Humming-Birds, but no additional information as to their habits and manners. A few years later (between 1829 and 1833) appeared M. Lesson's well-known works, the 'Histoire Naturelle des Oiseaux-Mouches,' 'Histoire Naturelle des Colibris,' and 'Les Trochilidées,'—publications which added considerably to our previous knowledge of the group, although they enumerate no more than 110 species. How little progress, then, had been made towards an intimate acquaintance with these lovely birds between the date of the twelfth edition of the 'Systema Naturæ' and that of the last-named publications, a period of more than seventy years!

If the illustrious Humboldt paid no very marked attention to the *Trochilidæ*, he must have noticed many of the fine species lately brought to light; and it is therefore somewhat surprising that he should have been so remarkably silent respecting them when writing the 'Personal Narrative' of his travels in the new world. It is to him and to his associate Bonpland, however, that I consider we are indebted for our acquaintance with many of them; for the perusal of the interesting account of their enterprising travels has doubtless created a desire in others to follow in their footsteps. Thus succeeding travellers, who have not been slow to perceive how wonderfully different are the productions of the great Andean ranges from those of the other parts of South America have ever been active in forming and transmitting to Europe collections in nearly every department of science, and no objects have been more assiduously sought for than the flying gems which constantly greeted them at every turn and must have been always before their eyes. Among the most eminent travellers who have succeeded Humboldt are D'Orbigny, Schomburgk, Tschudi, Castelnau, Burmeister, and others, who, with more recent but less known explorers, have added so largely to our knowledge of the *Trochilidæ*. Both Frenchmen and Belgians have proceeded to South America to procure supplies of these birds; and dealers from those countries have established themselves in some of the cities of that part of the world for the like purpose. From Sta. Fé de Bogota alone many thousands of skins are annually sent to London and Paris, and sold as ornaments for the drawing-room and for scientific purposes. The Indians readily learn the art of skinning and preserving, and, as a certain amount of emolument attends the collecting of these objects, they often traverse great distances to procure them; districts more than a hundred miles on either side of Bogota are strictly searched; and hence it is that from these places alone we receive not less than seventy species of this family of birds. In like manner the residents of many parts of Brazil employ their slaves in collecting, skinning, and preserving them for the European market; and many thousands are annually sent from Rio de Janeiro, Bahia, and Pernambuco. They also supply the inmates of the convents with many of the more richly coloured species for the manufacture of artificial-feather flowers. How numerous, then, must these birds be in their native wilds, and how wonderfully must they keep in check the peculiar kind of insect life upon which they principally feed! which is, doubtless, one of the objects for which they were designed. After these few cursory remarks I proceed to give a general history of the group, the range and distribution of the species, and such additional information as I have acquired during the course of my labours.

"The first mention which is made of the Humming-Birds," says M. Lesson, "in the narratives of the adventurers who proceeded to America, not with the design of studying its natural productions, but for the discovery of gold, dates from 1558, and is to be found in 'Les Singularités de la France Antarctique' (Brazil) of André Thevet and Jean de Lery, companions of La Villegaignon, who attempted in 1555 to found a French colony there; but these superficial accounts would not have unfolded their natural history, had not the old naturalists who published their observations at the commencement of the seventeenth century taken care to make them better known; and we find some good accounts of them in the voluminous compilation of Nieremberg, in the collection of fragments from the great works of Hernandez or Fernandez, and in those of Piso. Ximenez, Acosta, Gomara, Marcgrave, Garcilasso, and Dutertre often mention these birds, but their remarks are so superficial that it would be of little use to quote them now. Towards the end of the same century Sir Hans Sloane, Catesby, Edwards, Brown, Father

Labat, Plumier, Louis Feuillée, and Rochefort gave tolerably complete figures and descriptions of some of the species; but it was not until the commencement of the eighteenth century that we became better acquainted with their natural history."

It will be seen that little was really known respecting the Humming-Birds even at the end of the career of the great Linnæus. From Captain Cook both Pennant and Linnæus became aware that a species was found as far north as Nootka Sound, while every voyager to the eastern shores of North America brought tidings of its representative in the *Trochilus colubris*. Jamaica, St. Domingo, and the smaller islands of the West Indies, furnished a fair quota in the species inhabiting those countries; and correspondents were speedily established by Sloane, Brown, Edwards and Catesby in Hispaniola, Demerara, and Brazil. Of all these countries the Humming-Birds and other zoological productions were then but partially, and only partially, known. The great primeval forests of Brazil, the vast palm-covered districts of the deltas of the Amazon and the Orinoco, the fertile flats and savannahs of Demerara, the luxuriant and beautiful region of Xalapa (the country of perpetual spring) and other parts of Mexico, were literally untrodden ground by the ornithological collector. Up to this time the vast provinces of the New World had only been skirted; all within was virgin land, wherein even the explorer had scarcely placed a foot, and where the only human inhabitants were the wild children of nature—the Botacudos and other tribes of South American Indians. If the country glanced at in the foregoing remarks had provided the naturalists of the days of Linnæus with ample materials for study and investigation, how much greater would have been their amazement and delight had they been acquainted with the hidden treasures of the great Andean ranges, which stretch along the entire country, from the Rocky Mountains on the north to near Cape Horn on the south. Along the whole line of this great backbone, as it were, of America, at remarkably short intervals, occur species of this family of birds of the greatest beauty and interest, which are not only specifically but generically distinct from each other. Whole groups of them, remarkable for their singularity, have become known to us from the inquiries and explorations of later travellers; and abundant as the species may be towards the northern and southern portions of the great chain of mountains, they vastly increase as we approach the equator. These equatorial regions teem with species, and even genera, which are not found elsewhere. Between the snowline of the summits of the towering volcanoes and their bases, many zones of temperature occur, each of which has its own especial animal and vegetable life. The alpine region has its particular flora, accompanied by insects especially adapted to such situations; and attendant upon these are peculiar forms of Humming-Birds, which never descend to the hot valleys, and scarcely even to the cooler and more temperate paramos. Many of the highest cones of extinct and of existing volcanos have their own faunas and floras: even in the interior walls of ancient craters, wherever vegetation has gained a footing, some species of Humming-Birds have there, and there only, been as yet discovered. It is the exploration of such situations that has led to the acquisition of so many additional species of this family of birds, which now reach to nearly 400 in number.

It might be thought by some persons that 400 species of birds so diminutive in size, and of one family, could scarcely be distinguished from each other; but any one who studies the subject, will soon perceive that such is not the case. Even the females, which assimilate more closely to each other than the males, can be separated with perfect certainty; nay, even a tail-feather will be sufficient for a person well versed in the subject to say to what genus and species the bird from which it has been taken belongs. I mention this fact to show that what we designate a species has really distinctive and constant characters; and in the whole of my experience, with many thousands of Humming-Birds passing through my hands, I have never observed an instance of any variation which would lead me to suppose that it was the result of a union of two species. I write this without bias, one way or the other, as to the question of the origin of species. I am desirous of representing nature in her wonderful ways as she presents herself to my attention at the close of my work, after a period of twelve years of incessant labour, and not less than twenty years of interesting study. I am, of course, here speaking of the special object of my own studies—the Humming-Birds.

It is somewhat remarkable that any persons living in the present enlightened age should persist in asserting that Humming-Birds are found in India and Africa. Yet there are many who believe that such is the case. Even in a work but recently published it is stated that Humming-Birds and Toucans are both found in the last-mentioned country; and I was once brought into a rather stormy altercation with a gentleman who asserted that the Humming-Bird was found in England, and that he had seen it fly in Devonshire. Now the object seen in Devonshire was the insect called the Humming-Bird Moth, *Macroglossa stellarum*; and the birds supposed to belong to this family by residents and travellers in India and Africa are of a totally different group—the *Nectariniidæ* or Sun-Birds. These latter birds have no relationship to the *Trochilidæ*; they are not even representatives of them in the countries alluded to; and their only points of resemblance consist in their diminutive size and the showy character of their plumage. Let it be understood, then, once for all, that the Humming-Birds are confined to America and its islands (that is, the West Indies in the Atlantic, and Chiloe and Juan Fernandez in the Pacific; none have as yet been found in the Galapagos). The *Selasphorus rufus* goes as far north as Sitka. Kotzebue informs us that it is found in summer as high as the sixty-first parallel on the Pacific coast; while, on the antartic end of the continent, Captain King observed the *Eustephanus galeritus* flitting about among the Fuchsias of Tierra del Fuego in a snow-storm. Both these species, however, are migrants,—the northern bird retiring, as autumn approaches, to the more temperate climate of Mexico, while the other wends its way up to the warmer regions of Bolivia and Peru. The migration of these birds is of course performed at directly opposite periods. Both the *Selasphorus rufus* and the *Trochilus colubris* spend the summer in high northern latitudes; but the former always proceeds along the western, and the latter along the eastern parts of the country: the *T. colubris* even extends its range as far as the fifty-seventh parallel, where it was observed by Sir John Richardson. Although these and some other species pass over vast extents of country, I do not believe that they are capable of long-continued flights: that is, I question their power of crossing seas or more than from one island to another; for although we know that the two birds above-mentioned pass over many degrees of latitude in their migrations, I believe that these journeys are performed in a series of comparatively short stages, and always by land, and that the whole of their movements are more or less influenced by the progress of the sun north or south as the case may be.

North America, then, may be said to have two Humming-Birds— a western and an eastern species. It is true that Audubon has mentioned two others in his great work—the *Lampornis Mango* and *Calypte Annae*—and states that the former was found at Key West in East Florida. Since then, however, I believe no other example has been discovered there; and one can scarcely understand the occurrence of the bird in that part of America, since it is a native of countries and islands lying so much further south.

Leaving North America, and proceeding south, we begin to meet with several other species, which rarely extend their range to the north—viz. the *Calypte Annae*, *C. Costae*, *Selasphorus platycercus*, *Trochilus Alexandri*, and *Calothorax Calliope*. These birds are also migratory, but their range is much less extensive than that of the two species previously mentioned. As we advance in this direction, Humming-Birds become extremely numerous, and, as regards genera and species, continue to increase in the more southern country of Guatemala, where every variety of climate is to be found. The forest-clad mountains of Vera Paz appear to afford a winter retreat to many of the northern species, as the regions contiguous to the Atlas range in Africa do to the numerous little warblers of this country and the continent of Europe. Besides these migrants, Guatemala, Honduras, and Costa Rica have species which are either stationary or merely change their quarters in accordance with the flowering-season of the trees on which they seek their food, moving east and west or *vice versâ* according to circumstances. The countries further south, or those lying between Guatemala and Panama, appear to have a bird-fauna almost peculiar to themselves; for it is seldom that the species inhabiting Costa Rica and Veragua extend their range to the northward, neither are they often found in the more southern country of New Granada.

It is in the last-mentioned country—New Granada—that some of the finest of the Trochilidæ are found,—its towering mountains having species peculiar to themselves, while its extensive paramos are tenanted by forms not found elsewhere. On the principal ranges of the Andes, species exist which do not occur on the lower elevations situated more to the eastward. These ranges are the sources of numerous rivers, some of which have a northerly course —such as the Atrato, Cauca, and the great Magdalena, which debouch into the Caribbean Sea—and the river Zulia, which empties itself into the Lake of Maracaybo. Some of the very finest species yet discovered were collected near the town of Pamplona, which is situated on the banks of the last-mentioned river. The country round Antioquia, situated on the lower, and Popayan on the upper part of the Cauca, appear also to be very rich in natural productions, and particularly so in Humming-Birds. It is, however, on the paramos which surround Bogota, and on the luxuriantly clad sides of the valleys through which flows the main stream of the Magdalena, that the greatest number of species have been discovered. Bogota, the capital of this district, has for a long time been the centre whence collections have been transmitted to Europe and the United States. The Indians have been initiated into the modes of preparing these lovely objects; and as gain and excitement have thus gone hand in hand, this part of America may be said to have been thoroughly ransacked, and I expect that but few novelties remain to be discovered therein. Now as most of the productions that have yet reached us from Antioquia and Pamplona, two districts lying in about the same parallel of latitude on either side the great valley of the Magdalena, are quite distinct and different from those of Bogota, we may safely infer that, if they were as closely searched, many new species would be found. The country of the Caraccas and Cumana have Humming-Birds which partake less of the characters of the mountain species, and assimilate more closely to those of the Guianas, and Northern Brazil. It will be seen, I think, from what I have here said, that the species of Humming-Birds increase in numbers as we proceed towards the equator; that most of them are confined to countries having peculiar physical characters; and that those of New Granada differ considerably from the Humming-Birds of Veragua, Costa Rica, and Guatemala. I have observed an equally marked difference in the species which inhabit the high lands giving rise to the rivers which run eastward; I mean the many tributaries of the Napo, the Caqueta or Japura, and the Amazon.

From the eastern side of Chimborazo flow many streams which ultimately find their way into the Amazon; and however numerous the species found in the elevated districts of New Granada may be, I believe that when the dense and luxuriant forests bordering these well-watered lands are fully investigated, the species inhabiting them will be found far to exceed in number those of every other district. Even the snowy Chimborazo may be said to be inhabited by Humming-Birds: certain it is that the *Oreotrochilus Chimborazo* lives upon it just below the line of perpetual congelation, some of my specimens of this bird killed by M. Bourcier bearing on the attached labels an elevation of 16,000 feet; and Mr. Fraser, I believe, killed others in an equally elevated region. Here, then, is a bird which encounters the cold blasts of these lofty situations with impunity, dwelling in a world of almost perpetual sleet, hail, and rain, and there feeding upon the insects which resort to the *Chuquiraga insignis* and other flowering plants peculiar to the situation. These truly alpine birds have always a great charm with me; and as the species just mentioned is especially beautiful, it is of course a great favourite. Besides Chimborazo, there exist many other cones of but little less elevation, such as Pichincha, Cotopaxi, and Cayambe, which, strange to say, are reported to be frequented by species peculiar to each; and if this be the case, how many other summits yet untrodden may reveal others at present unknown to us? Now what I have said with regard to the gradual increase of Humming-Bird life from the north to the equator may be equally said of their increase towards the same line from the south. The species there found, although quite different from those of the north, perform precisely the same functions, are subject to the same migratory movements, &c.

To the southward of the equator, however, the species appear to be far less numerous. And it could not be expected but that such would be the case when we consider the particular character of the country,—the dry and sterile plains of Peru, the extensive pampas of La Plata, &c., being all unsuited to insect and therefore to Humming-Bird life, and a diminution in their numbers the natural result. But the paucity in numbers would seem to be compensated in the beauty of the individuals. Peru and Bolivia are the cradles of the splendid comet-tailed species of the genus *Cometes*, the *Lesbiæ*, *Diphogenæ*, the delicate birds known as *Thaumasturæ*, &c. These countries produce also the largest Humming-Bird yet known, the *Patagona gigas*, which with an *Oreotrochilus* and a *Eustephanus* are all the species known to me from the lengthened country of Chili. The little island called Chiloe, characterized by great humidity, is inhabited by the common Chilian species last mentioned; while the celebrated island of Juan Fernandez, over 300 miles from the mainland, is tenanted by three kinds, of which two are so distinct from all others known, that they cannot for a moment be confounded with any of them. The three species, in fact, which people this solitary spot in the wide Pacific are very different from each other; and I may mention that nothing like a cross or intermixture has ever been observed, an event that might have been expected to occur here, if ever it does among animals living in a state of nature. Strange to say, these beautiful creatures are almost the only examples of bird life existing on this remarkable island. The knowledge of the existence of these lovely flying gems gives an additional zest to the interest attached to the scene of the principal events in Defoe's charming tale.

In the foregoing pages I have glanced at the species of Humming-Birds inhabiting the great range of mountains running north and south through many degrees of latitude on both sides of the equator. Whole genera of the Trochilidæ are found there, and there alone. In the highlands of Mexico, among others we find the peculiar genera *Delattria*, *Selasphorus*, and *Calypte*. On crossing the ribbon-like strip of land called the Isthmus of Panama, we enter upon a region of highlands bearing the genera *Oxypogon, Lafresnaya, Bourcieria, Doriferа, Helianthea, Heliangelus, Eriocnemis, Lesbia, Cynanthus, Aglæactis, Metallura, Ramphomicron*, and many others, none of which are found in the less-elevated countries of Brazil, the Guianas, or the West Indian Islands. It is true that these countries, particularly Brazil, possess forms of Humming-Birds which are now and then feebly represented in the Andes; but these cases are quite exceptional. When we leave the Andes we bid adieu to the finest, the largest, and the most gorgeously attired species. Other beautiful kinds do here and there exist in Brazil, such as the *Chrysolampis moschitus*, the *Topaza pella*, and the *Lophornithes*; but the greater number are comparatively small and inconspicuous. Of the members of the genus *Phaëthornis*, a group of Humming-Birds, popularly known by the name of Hermits, from their frequenting the darkest and most retired parts of the forest, three-fourths are natives of Brazil. The great forest-covered delta of the Amazon, where palms are numerous, seems to be particularly unfavourable to the *Trochilidæ*, since from Para to Ega there are scarcely ten species of the family to be met with.

In this cursory glance at the distribution of this family of birds, those frequenting the West Indian Islands have yet to be noticed; and here not only do we find some peculiar to those islands as a whole, but in each of them, with but very few exceptions, there are species and even genera which are not found in the Andes, the other islands, or the more contiguous flat parts of the South American Continent. Cuba has at least three, one of which is a most lovely little bird. The principal island of the Bahaman group is in like manner favoured with a charming *Calothorax*, which Dr. Bryant tells us flies in great numbers round the town of Nassau; yet the bird does not, I believe, inhabit any of the other islands or the mainland.

Jamaica possesses three, which are all quite distinct, and so widely different from every other, that it is a perfect mystery to the naturalist how they first obtained a footing there. Nothing like interbreeding between two species appears to occur in this island; if such were the case, we could not but be aware of the fact, since we have not only been for many years in the habit of receiving hundreds of birds from Jamaica, but this island has had the advantage of a naturalist, Mr. Gosse, who has most closely observed the birds resident there. St. Domingo has two species, differing from those of Jamaica. This law with respect to the Humming-Bird inhabitants of the West Indian and Leeward Islands, is equally carried out in the necklace-like string of the Windwards; but when we arrive at the island of Trinidad, the species become much more numerous and partake of the character of those which inhabit the mainland—the opposite shores of Venezuela.

It may be asked, what is our present knowledge of the existing species of Humming-Birds, and if there may not be others to be discovered in the great primeval forests of the western and other parts of the vast continent of the new world. My reply is that, in all probability, many more than are known to us do exist, and that a very lengthened period must elapse before we shall acquire anything like a perfect knowledge of the group. Whatever I may have done towards the elucidation of the subject, I must only be regarded as a pioneer for those who, in future ages, will render our acquaintance with this family of birds so much more complete than it is at the present time.

The countries of South America whose productions are least known are Costa Rica, Veragua, Panama, the sea-bord between Carthagena and Guayaquil, the forests of La Paz and other parts of Bolivia, the whole of the eastern slopes of the Andes bordering Peru and Ecuador, and the western portion of Brazil. All these countries will doubtless furnish new kinds of Humming-Birds when the explorer shall extend his researches into their unknown recesses. We may feel fully convinced that such will be the case from the circumstance of single individuals in a youthful or imperfect state, which we cannot identify as belonging to any known species, occasionally occurring in the great collections sent from time to time to Europe. My own collection contains several examples of this kind, which will doubtless at some future day prove to belong to undescribed species. For more than twenty long years have I been sending the most earnest entreaties, accompanied with drawings, to my correspondents in Peru and Ecuador for additional examples of that truly wonderful bird the *Loddigesia mirabilis*. These entreaties have been backed by the offers of large sums of money to any person who would procure them; but up to the present moment no second example has been obtained. Probably the single individual killed by Mr. Matthews in the neighbourhood of Chachapoyas was one which had accidentally strayed beyond the area in which the species usually dwells, and which has not yet been discovered. That it may be a nocturnal bird has sometimes suggested itself to my mind, and that this may be the reason why it has not since been seen. Those of my readers who are not acquainted with this most wonderful member of the Trochilidæ will do well to refer to the plate, in which a correct representation of it is given by the masterly hand of Mr. Richter.

The preceding remarks must, I think, have given the reader a general idea of the countries inhabited by the members of the great family of Humming-Birds; it now becomes necessary to speak of their peculiar structure, and the place they appear to occupy in the Class Aves. By systematists they have been bandied about from one group to another: by some they have been associated with the Sun-Birds (*Nectariniæ*); by others with the *Cypselinæ*, *Picinæ*, *Sittinæ*, *Certhinæ*, &c.

In Brisson's arrangement, published in 1760, they constitute with the Creepers his twelfth order. By Linnæus in 1766, and Latham in 1790, they were placed in the class *Picæ*, together with the Creepers, Hoopoes, &c. In like manner they are associated with the same birds in the fourteenth order of Lacépède's arrangement, published in 1799. In Duméril's classification, proposed in 1806, they form part of his second order—Passerine Birds—and are associated with Kingfishers, Todies, Nuthatches, Bee-Eaters, Creepers, &c. They form a distinct family of the second Order, *Ambulatores*, in the arrangement of Illiger published in 1811. They also constitute a distinct family by themselves of the Tenuirostral Division of the order *Passeres* in Cuvier's system of 1817. By Vieillot, whose arrangement was published about the same time, they form part of the twenty-second family *Sylvicolæ*, and are associated with Creepers, Sun-Birds, and Honey-eaters. By Temminck, in the second edition of his 'Manuel d'Ornithologie,' published in 1820, they were placed, together with the Creepers, Sun-Birds, Hoopoes, &c., in his sixth Order, *Anisodactyli*. In De Blainville's arrangement, which appeared in the years in 1815, 1821, and 1822, they form a separate family of the *Saltatores*, with the Kingfishers preceding, and the Crows following them. Vigors, in 1825, made them a distinct family of his second Order, *Insessores*,—the preceding family being composed of the Sun-Birds, and the succeeding one of the *Promeropidæ*. Latreille in the same year placed them in the fourth family *Tenuirostres* of the second Order or Passerine Birds, along with the Hoopoes, Promerops, Sun-Birds, &c. Lesson, in 1828, made them the eighth family of the *Insessores*, and associated them with the Sun-Birds, Creepers, &c. By Boié they were divided in the 'Isis' for 1831 into eleven genera, viz. *Hellatrix, Callipkloæ, Glaucis, Anthracoraæ, Heliactin, Hylocharis, Basilinna, Chrysolampis, Heliothrix, Smaragdites,* and *Eulampis*. Swainson, in 1837, constituted them the third family of the *Tenuirostres*, with the Sun-Birds preceding, and the Promeropidæ and Hoopoes succeeding them. In Mr. G. R. Gray's 'List of the Genera of Birds,' published in 1841, and in his great work 'On the Genera of Birds,' completed in 1850, they form the third family of the Tenuirostres. In the 'Conspectus Systematis Ornithologiæ' of Prince Charles Lucien Bonaparte, given to the world a few years before his lamented death, they form Stirps 17 *suspensi*, of his second Order PASSERES; and Tribe *Volucres*, with the Hoopoes and Promerops placed before, and the Swifts and Swallows after them. In his 'Conspectus Generum Avium' they form the eleventh family of the *Insessores*, with the Swifts preceding them, and are succeeded by the *Phytotomidæ* or Plant-Cutters. In his "Conspectus Trochilorum," published in the 'Revue et Magasin de Zoologie' for May 1854, they form the seventy-second family of his Passerine Birds. In Dr. Reichenbach's arrangement, in Cabanis's 'Journal für Ornithologie' for 1853, they are fancifully divided into groups of Fairies, Elfs, Gnomes, Sylphs, &c.; and in his 'Trochilinarum Enumeratio' he places these birds between the true Creepers on the one hand, and the Hoopoes on the other. By Cabanis, the latest writer on the subject, they are placed with the Swifts and Goatsuckers, in his 3rd Order *Strisores* and Tribe *Macrochires*.

Ornithologists of the present day consider them to be more intimately allied to the true Swifts than to any other group of birds. This view of the subject is supported by the fact of the Humming-Birds, like the Swifts, having most ample wings, vast powers of flight, and a bony structure very closely assimilating: and this alliance is still further exemplified in some parts of their nidification, the number and colour of their eggs, &c. It is not to be expected that, with this subject before me for so many years, I should have been inattentive to the consideration of the place these birds should occupy in our attempts at a natural arrangement; and while I admit that they are somewhat allied to the Swifts, they are so essentially distinct from these and all other birds, that they might be separated into a distinct Order with quite as much, if not greater, propriety as the Pigeons when considered in relation to the Gallinaceous Birds. They have certain characters, dispositions, and modes of life which are not to be noticed in any other group of birds: their cylindrical bills, double-tubed tongues, enormously developed sternums, and corresponding pectoral muscles, rigid primaries, the first of which is the longest, and their diminutive feet separate them from all others. In the Swifts and Fissirostral birds generally the sexes are alike in outward appearance; in the Humming-Birds they are in nearly every instance totally different in their colouring: in the former the young assume the livery of the adult before they leave the nest, while the contrary is the case with the Humming-Birds. How different, too, is the texture of the luminous feathers with which they are clothed; and vastly diversified in form as the tail is in the various genera, the number of feathers in the whole of them is invariably ten. In their disposition they are unlike birds, and approach more nearly to insects. Many of the species fearlessly approach almost within reach of the hand; and if they enter an open window, as curiosity may lead them to do, they may be chased and battled with round the apartment until they fall exhausted; and if then taken up by the hand, they almost immediately feed upon any sweet, or pump up any fluid, that may be offered them, without betraying either fear or resentment at their previous treatment. A *Trochilus colubris*, captured for me by some friends at Washington (Baron Osten Sacken, Mr. Odo Russell, and his brother Mr. Arthur Russell), immediately afterwards partook of some saccharine food that was presented to it, and in two hours it pumped the fluid out of a little bottle whenever I offered it; and in this way it lived with me a constant companion for several days, travelling in a little thin gauzy bag distended by a slender piece of whalebone, and suspended to a button of my coat. It was only necessary for me to take the little bottle from my pocket to induce it to thrust its spiny bill through the gauze, protrude its lengthened tongue down the neck of the bottle, and pump up the fluid until it was satiated; it would then retire to the bottom of its little home, preen its wing- and tail-feathers, and seem quite content.

The specimens I brought alive to this country were as docile and fearless as a great moth or any other insect would be under similar treatment. The little cage in which they lived was twelve inches long, by seven inches wide, and eight inches high. In this was placed a diminutive branch of a tree, and suspended to the side a glass phial which I daily supplied with saccharine matter in the form of sugar or honey and water, with the addition of the yelk of an unboiled egg. Upon this food they appeared to thrive and be happy during the voyage along the sea-bord of America and across the Atlantic, until they arrived within the influence of the climate of Europe. Off the western part of Ireland symptoms of drooping unmistakeably exhibited themselves; but although they never fully rallied, I, as before stated, succeeded in bringing one of them alive to London, where it died on the second day after its arrival at my house. The vessel in which I made the passage took a northerly course, which carried us over the banks of Newfoundland; and although the cold was rather severe during part of the time, the only effect it appeared to have upon my little pets was to induce a kind of torpidity, from which, however, they were readily aroused by placing them in the sunshine, or in some warm situation, such as before a fire, in the bosom, &c. I do assure my readers that I have seen these birds cold and stiff, and to all appearance dead; and that from this state they were readily restored with a little attention and removal into light and heat, when they would "perk up," flutter their little wings, and feast away upon their usual food as if in the best state of health.

How wonderful must be the mechanism which sets in motion and sustains for so lengthened a time the vibratory movements of a Humming-Bird's wings! To me their action appeared unlike any thing of the kind I had ever seen before, and strongly reminded me of a piece of machinery acted upon by a powerful spring. I was particularly struck by this peculiarity in the flight, as it was exactly the opposite of what I expected. The bird does not usually glide through the air with the quick darting flight of a swallow or swift, but continues tremulously moving its wings while passing from flower to flower, or when taking a more distant flight over a high tree or across a river. When poised before any object, this action is so rapidly performed that it is impossible for the eye to follow each stroke, and a hazy semicircle of indistinctness on each side of the bird is all that is perceptible. "The wind produced by the wings of these little birds," says Mr. Salvin, "appears to be very considerable; for I noticed that while an example of *Cyanomyia cyanocephala* which had flown into the room was hovering over a large piece of wool, the entire surface of the wool was violently agitated." Although many short intermissions of rest are taken during the day, the bird may be said to live in air—an element in which it performs every kind of evolution with the utmost ease, frequently rising perpendicularly, flying backward, pirouetting or dancing off, as it were, from place to place, or from one part of a tree to another, sometimes descending, at others ascending; it often mounts up above the towering trees, and then shoots off like a little meteor at a right angle; at other times it quietly buzzes away among the little flowers near the ground; at one moment it is poised over a diminutive weed, at the next it is seen at a distance of forty yards, whither it has vanished with the quickness of thought. During the heat of the day the shady retreats beneath the trees are very frequently visited; in the morning and evening the sunny banks, the verandahs, and other exposed situations are more frequently resorted to.

The foregoing remarks are from personal observation of the habits of *Trochilus colubris*; and I have been informed by Mr. Salvin and others that a similar action characterizes most of the species. I believe, however, that those members of the Trochilidæ which are furnished with more ample wings, such as the species of the genera *Aglæactis, Ramphomicron, Pterophanes,* and *Patagona,* have a very different mode of flight, move their wings with diminished rapidity, and pass much more slowly through the air. Mr. Darwin, when speaking of the *Patagona gigas,* says, "Like others of the family, it moves from place to place with a rapidity which may be compared to that of *Syrphus* among Diptera, and *Sphinx* among Moths; but whilst hovering over a flower it flaps its wings with a very slow and powerful movement, totally different from that vibratory one, common to most of the species, which produces the humming noise. I never saw any other bird, where the force of its wings appeared (as in a butterfly) so powerful in proportion to the weight of its body. When hovering by a flower, its tail is constantly expanded and shut like a fan, the body being kept in a nearly vertical position. This action appears to steady and support the bird, between the slow movements of its wings."

In the intervals of flight, I believe that they not only rest in the ordinary way, but even pass some time in sleep; at least I found that this was the case with my living birds, and that from this state of partial torpor they were not easily aroused. In the morning and evening they were far more animated than at any other period of the day; and they would even perform their buzzing evolutions round their cage, and sip from their little bottle in the night-time, if a light was brought into the room. They usually sat in a moping position, with the bill in a line with the body, or slightly elevated, after the manner of the Kingfishers. I never saw them hang by their feet and sleep with their heads downwards—a position which I have been informed is sometimes assumed by Humming-Birds.

When we have compared the wings of *Calliphlox Amethystinus* with those of *Patagona gigas,* we have noticed the two extremes of development in these organs, but many intermediate forms exist, and each modification has doubtless an influence on the mode and power of flight. I cannot leave the subject of the wings without alluding to the extraordinary development of the shafts of the primaries in the *Campylopteri.* The great dilatation of these feathers would lead one to suppose that they have an influence on the aërial movements of the birds; but, strange to say, this remarkable feature only occurs in the males; the females are entirely destitute of it. It might naturally be supposed that such a modification of so important an organ must be formed with an especial object. What, then, can be the particular use of the broad dilated shafts of these singularly and apparently awkwardly shaped wings? Generally the primaries and secondaries are of a sombre and uniform hue, while the shoulders or wing-coverts, in most instances, are of the same colour as the other parts of the body. There are, however, a few, but a very few exceptions to the rule; and I may mention the *Eulampis jugularis* and *Pterophanes Temmincki* as instances in point: both these birds have luminous wings, and must form very striking objects during flight; and, as I believe colour is seldom given without the intention of its being exhibited, there is doubtless something peculiar in the economy of these birds. The primaries and secondaries are in some instances stiff and rigid, while in others they are soft and yielding; some are broad, others narrow; they are always the same in number, and the first quill is constantly the longest, except in *Polytmus cephalater*, where the second exceeds the first in length.

When we turn to the bill, we find this organ to be greatly diversified in form, and that each of these variations appears to be specially adapted for some given purpose; indeed, I have never seen the law of adaptation more beautifully exemplified than in the multiplied forms exhibited in the bills of the members of the various genera of this family of birds. A certain generic character runs through the whole of them; the gape in all cases is very small, and whether the bill be curved or straight, the upper mandible overlaps the under one on both sides, and thus forms an admirable protection for the delicate double-tubed tongue. If we examine the extraordinarily lengthened bill of *Docimastes ensifer* and the short feeble bill of the *Lesbia Gouldi*, we see the extremes as regards the length of this organ; and we are not less astonished at the functions they are both intended to perform. The bill of the *D. ensifer*, which is nearly six inches long, and which contains a tongue capable of being protruded nearly as far beyond its tip, is most admirably fitted for the exploration of the lengthened and pendent corollas of the *Brugmansiæ*; while the short-billed *Lesbiæ* cling to the upper portion of those flowers, pierce their bases, and with the delicate feelers at the extremities of the tongue, readily secure the insects which there abound. I have been assured by M. Bourcier that this is really a practice of the bird, and that it frequently resorts to this device for the purpose of gaining its insect food; but I suspect that, besides exploring the stalwart *Brugmansiæ*, a more delicate flora is the object for which its bill is especially formed. In no part of America are so many tubular-flowered plants as among the Andes, and the greater number of the Humming-Birds found there have straight and lengthened bills, such as the members of the genera *Helianthea*, *Bourcieria*, *Cœligena*, etc. The arched bills of the *Phaëthornithes* are admirably adapted for securing the insects which resort to the leaves of trees, and upon which these birds are said to exist. But how much are we astonished, when we examine the bill of *Eutoxeres*! and find this organ curved downwards beyond the extent of a semicircle, a form beautifully adapted for exploring the scale-covered stems of the larger palms.

Let us turn to another genus of this group—*Grypus*. Here the bill is not only armed with a strong hook at the end of the mandibles, but with a row of numerous and thickly set teeth. The *G. nævius* is said to frequent the borders of the great forests, and to gain its food from among the interstices of the bark of the palm trees. Both this bird and the *Eutoxeres*, as well as the *Phaëthornithes*, are said (and, I believe, with truth) to feed principally upon spiders; and we know that these are the food of the *Grypus*. All the members of the genus *Ramphomicron* are said to feed on insects which inhabit the

c 2

alpine Floræ; and their bill is well suited to the capture of the minute insects found in those elevated regions. In some instances the bill is perfectly wedge-shaped, as in *Heliothrix*; while in others it suddenly turns upwards, as in *Avocettula*. These forms are also adapted for some special purpose, of which, however, at present we are ignorant. Besides these, there are others whose bills approach somewhat to the form of the flycatchers, as the *Aithurus*. This bird we know frequently seizes insects on the wing; and so doubtless do many of the others. It will have been seen that all these forms of bill are well suited for the capture of insects; and, as might be supposed, insects constitute the principal food of the Humming-Bird; but that liquid honey, the pollen and other saccharine parts of flowers are also partaken of is evident from the double tubular tongue with which all the species are provided. Besides this they readily and greedily accept this kind of food when offered to them in a state of captivity, or when the corollas of a bouquet of flowers placed in a window are filled with sugar to entice them to approach; and from my own experience I know that they have been kept in captivity for several months upon this kind of food.

Connected intimately with the mode of flight is the form and structure of the tail, and in no group of birds is this organ more varied; in some species it is four times the length of the body, in others it is so extremely short as to be entirely hidden by the coverts. As cases in point I may mention *Lesbia Amaryllis* and *Calothorax micrurus*. Every Humming-Bird, however, has ten tail-feathers, and no more. I am aware that this number is not apparent in some of the smaller fork-tailed species, the two centre-feathers being so exceedingly minute as to be almost obsolete; but if a careful examination be made, that number will be found. I may instance *Thaumastura Coræ*, *T. enicura*, and *Calothorax Fanniæ*.

The tail appears to be, and doubtless is, a very important organ in all the aërial movements of the Trochilidæ; and accordingly we find very great variations in its form among the many different genera of which the family is composed. In *Cometes* and *Lesbia*, the forked character is carried to its maximum, while its minimum is seen in *Calothorax*, *Acestrura*, and the allied groups. The tails of all the members of the two former and many other genera are of this form; while in others it is only seen in a single species of a group, all the other members of which have rounded, square, or cuneate tails. As a case in point I may cite *Eupetomena hirundinacea*, among the Campylopteri, which may be regarded as the aërial type of its own particular group. Next to this we may notice the species with feathers terminating in spatules, such as *Loddigesia*, *Spathura*, etc. I was informed by the late Mr. Dyson that the flight of these birds presents a marked difference from that of other Humming-Birds, and that their appearance in the air is most singular,—the tail being not only constantly opened and shut, but the spatules always in motion, particularly when the bird is poising over a flower; and if this be really true, what an extraordinary appearance must the *Loddigesia mirabilis* present during its evolutions! But we cannot attempt to describe it; the discovery of a second example, and the peculiarity of its flight, must be left for future historians to make known to us.

In some few instances, such as *Julianyia typica* and *Campylopterus Pampa*, the tails are cuneate; but this form is quite exceptional if we exclude the *Phaëthornithes* and *Eutoxeres*, in which this is the prevailing form. Besides the groups with forked or cuneate tails, there are others in which this organ is square or rounded, as in the *Florisugæ* and *Metalluræ*. The reverse of the spatulate form occurs in some species, such as the members of the genus *Gouldia*, in which the tip of the outer tail-feathers terminates in thread-like filaments. The citation of one more will be sufficient to show how widely different is the form of this organ among the various genera. The outer feathers of the *Oreotrochili* are narrow, rigid, and turned inwards: this calliper-like form one might suppose would assist, in combination with the lengthened hind toe and claw, in supporting the bird on the sides of rocks; and we find that this is really the case; for Mr. Fraser informs me that he has seen several of the *Oreotrochilus Pichincha* clinging, half-benumbed with cold, on a ledge of rocks during one of the frequent snow-storms which occur on Pichincha. Quinarians would pronounce this to be the scansorial type among Humming-Birds. Now I think we may fairly infer that many of the other structures above alluded to are equally adapted for some peculiar purpose; yet there must be exceptions to this hypothesis, since the structure of the caudal feathers is in many instances totally different in the two sexes of the same species.

Nothing has yet been said respecting the legs and feet. Diminutive as they are, they will be found to be very diversified. In some instances the tarsi are bare, in others they are thickly clothed, as in the *Eriocnemides*; in some the toes are very diminutive, and are furnished with equally small, rounded nails; in others all the toes, particularly the hinder one, are greatly developed and armed with long, curved, and extremely sharp, spine-like claws. This latter form is admirably adapted for clinging to the petals of flowers—a habit common to many members of the family, which not only settle upon, but thrust their spiny bills through the bell-shaped flowers. The power these little birds possess of clinging to the branches is very remarkable: they hang on with their little feet and hooked claws like bats, with such pertinacity that I was often fearful of dislocating the legs of my living birds when attempting to remove them from their perch.

I may mention here, although somewhat out of place, that the skins of *Pterophanes Temminckii* have a strong musky smell, very similar to that exhaled by the Petrels. I consider this merely a coincidence; for although I am aware that many species of Humming-Birds fly close to the surface of water, they are merely hawking for insects among the aquatic plants peculiar to such situations.

It is the great diversity of forms in this family of birds which renders the study of them so very interesting. If these little objects were magnified to the size of Eagles, their structural differences would stand out in very bold relief, and the many marked generic distinctions they present would be far more clearly perceptible.

The preceding remarks have reference to such points of structure as may be considered to have an influence on the well-being of the birds. I shall now say a few words on those parts of the plumage which apparently are given for the purpose of ornament only,—the crests of *Cephalepis* and *Orthorhyncus*; the beards of *Ramphomicron* and *Oxypogon*; the ear-tufts of *Petasophora* and *Heliothrix*, the elegant appendages to the neck of the *Lophornithes*; the singular plume-like under tail-coverts of *Hypuroptila*, which in their structure and snowy whiteness strongly remind one of the corresponding feathers of the Marabou Stork, &c.

The members of most of the genera have certain parts of their plumage fantastically decorated; and in many instances most resplendent in colour. My own opinion is, that this gorgeous colouring of the Humming-Birds has been given for the mere purpose of ornament, and for no other purpose of special adaptation in their mode of life—in other words, that ornament and beauty merely as such was the end proposed—especially when we remember that the plumage of Humming-Birds seems to follow a general rule in the subordination and contrast with which the colours are arranged. These extraordinary developments are nearly always confined to the male, and are, doubtless, bestowed upon these little gems as a gorgeous train is given to the Peacock, beautiful markings to the Polyplectron, &c. I know of no others but the two species of the genus *Cephalepis* in which a single feather is made to serve the purpose of ornament. In all other instances the feathers are disposed in pairs, or in equal number on either side of the head or body, as the case may be; but in both these species the crest terminates in a single plume, which greatly adds to the elegance of the slender topping. How splendid are the spangles which deck the neck-plumes of the *Lophornithes*! and how well do the blue ear-tufts of the *Petasophoræ* harmonize with the surrounding green of the neck! The genera *Oxypogon* and *Ramphomicron* may be cited as singular instances of ornamentation; for they are both bearded and crested. Independently of these extra-developed portions of the plumage, certain parts of the body are gorgeously coloured; and here, again, some curious features are observable. In very many instances the crowns are truly resplendent, as in *Heliodoxa*; while in *Helianthea* the forehead only is decorated with a star brighter than Venus, the queen of planets.

All the members of the genus *Heliangelus* are remarkable for their beautiful gorgets, succeeded by a crescent of white separating it from the green of the under surface. Some species of the *Eriocnemides*, beside their thickly clothed tarsi, have rich and luminous upper tail coverts; while others, such as the *Eriocnemis Alinæ*, have the under tail-coverts unsurpassingly brilliant and beautiful. The members of the genus *Augastes* are conspicuous for the shining, metallike masks with which their faces are adorned; while, differing from all these, the *Aylaractines* have the lower part of their backs clothed in armour-like feathers, the brilliancy of which must be seen to be understood, but which, strange to say, is only apparent when viewed from behind; for if looked at in the direction of the feather, none of these hues are perceptible. Many more instances besides these might be mentioned; but a reference to the plates on which they are represented, or, still better, the birds themselves, will give a more correct idea of these remarkable colourings than can be conveyed by any description.

Before leaving the subject of extra development, I may mention that I often find it carried to a greater extent in some one species of a genus than in the others. I will give an example of what I here intend, by reference to what is observable in another family of birds, the *Trogonidæ*. Here the extra development of the upper tail-coverts which occurs in members of the genus *Pharomacrus*, commences in the *P. pavoninus*, increases in the *P. antisianus*, and extends beyond the tail in *P. auriceps*; but no species with upper tail-coverts of intermediate length between those of the last-mentioned species and the immensely long plumes of *P. paradiseus*, appears to exist. In like manner among the Andean Humming-Birds there is a tendency to a gradual increase in the length of the bill to the extent of two or two and a half inches; but no species has yet been seen in which that organ is intermediate between that length and the extraordinarily developed bill of *Docimastes*, which measures at least five inches. A similar fact is also observable with respect to the spatules in the *Spathura*.

Apart from development, I observe that in the Humming-Birds, as in some other groups to which I have paid particular attention, the species of one genus are much more numerous than those of others, and that, whenever this is the case, the genus usually comprises many closely allied species.

Among the most pleasing recollections of our youthful days is that of a birds' nest. Where is the person who has lived in the country and paid any attention to natural history, that does not recollect that of the Hedge-Sparrow (*Accentor modularis*) with its beautiful blue eggs; or has he ever ceased to wonder at the surprising construction of the nest of the Bottle-Tit (*Mecistura caudata*)? Their domestic architecture is indeed among the most interesting of the many singular features in the economy of birds; and how truly wonderful are some of the nests of the Humming-Birds! In form and size they vary as much as the different structures of the birds would lead us to expect, and a similar difference occurs in the situations in which they are placed. Some of these cradles are not larger than the half of a walnut-shell, and these coracle-shaped structures are among the neatest and most beautiful. The members of the genus *Trochilus* and their allies expend the greatest ingenuity, not so much in their construction as in the lavish decoration of their outer walls; with the utmost taste do these birds instinctively fasten thereon beautiful pieces of flat lichen, the larger pieces in the middle and the smaller on the part attached to the branch. It is a question among ornithologists whether these adornments are fixed on by a glutinous secretion from the bird or by the invisible webs of some of the smaller kinds of spiders; my own belief is, that the latter is the means employed. Now and then a pretty feather is intertwined or fastened to the outer side, the stem being always so placed that the feather stands out beyond the surface. These little cup-shaped nests are frequently placed on the bifurcation of the horizontal part of a branch near the ground, and at other times higher up towards the summit. Quite the reverse of this kind of nest are those built by the *Phaëthornithes*: these latter are generally very frail structures, woven round and attached to the side of a drooping palm-leaf, very frequently overhanging water. Such a nest is figured in my plate of *P. Eurynome*. Another of a similar form but of different materials is figured in the same volume in the plate illustrative of *P. Eremita*, with two young ones therein.

Other Humming-Birds suspend their nests to the sides of rocks. These are hammock-shaped in form, and are most ingeniously attached to the face of the rock by means of spiders' webs and the cottony materials of which they are sometimes built. Those made by the *Oreotrochili*, are very large, and composed of wool, llama hair, moss, and feathers; at the top of this great mass, of nearly the size of a child's head, is a little cup-shaped depression in which the eggs are deposited. Respecting the nest made by the *Oreotrochilus Pichincha*, my friend Professor Jameson, of Quito, writes, "On the first of the present month (November 1858), I visited the snowy mountain of Antisana in company with the American Minister. In the celebrated farm-house (about 13,500 feet above the sea) I found in one of the lower or ground apartments, unprovided with a door, several nests of *Oreotrochilus Pichincha*, one of which was attached to a straw rope suspended from the roof. I am quite certain as to the identity of the species, having shot one of the birds. The rest will be sent to you in my next parcel." See the figure of this nest given by Dr. Sclater in the 'Proceedings of the Zoological Society,' 1860, p. 80.

Some of the Humming-Birds, and perhaps this very species, are said to suspend their great nests by the middle from the fine hanging root of a tree, or a tendril; and should the nest, which is of a curved form and built of any coarse materials at hand, prove to be heavier on one side than the other, the higher side is weighted with a small stone or square piece of earth until an equilibrium is established and the eggs prevented from rolling out. If such powers so nearly approaching to that of reason should be doubted by some of my readers, I can assure them that one or more of these loaded nests are contained in the Loddigesian Collection; and one is at this moment before me, an examination of which will satisfy the most sceptical of the truth of this statement. Occasionally the old nests are repaired or built over the old one, two, three or more years in succession. Many other instances might be given to show that the nidification of the Humming-Birds is as singular as are the birds themselves. I believe that generally the eggs are two in number, but I also think it likely that some of the *Phaëthornithes*, or rather the members of the genus *Glaucis*, occasionally lay but one; for I have frequently seen only a single young bird in the nests sent to this country, and this single bird generally filled up the entire space of the frail structure, which, as I have before stated, is usually attached to the leaflet of a palm. The eggs are certainly large when we consider the tiny size of the birds which produce them; in shape they are oblong, nearly alike in form at both ends, and are probably of a pinkish hue before their contents are removed; after which they become of an opaque white, and so closely resemble bon-bons that they might easily be mistaken for them. The birds are said to produce two broods a year; and the period of incubation generally occupies about twelve or fourteen, or, according to Captain Lyon, eighteen days. This gentleman, when giving an account of some Humming-Birds whose hatching and education he sedulously watched, as the nest was made in a little orange-bush by the side of a frequented walk in his garden at Gongo Soco, in Brazil, states that the nest "was composed of the silky down of a plant, and covered with small flat species of yellow lichen. The first egg was laid January 26th, the second on the 28th; and two little creatures like bees made their appearance on the morning of February 14th. As the young increased in size, the mother built her nest higher and higher. The old bird sat very close during a continuance of heavy rain for several days and nights. The young remained blind until February 28th, and flew on the morning of March 7th, without previous practice, as strong and swiftly as the mother, taking their first dart from the nest to a tree about twenty yards distant."

Let me now mention one of the devices employed for the discovery of the nest of the Humming-Birds. Every observer who has written upon them has not failed to descant upon their boldness and pugnacity: not only do they attack birds of much larger size than themselves, but it is even asserted that they will tilt at the Eagle if he approaches within the precincts of the nest; nor is man exempt from their assaults, of which an amusing instance will be found in the extract from Lady Emmeline Stuart Wortley's 'Travels' given on a subsequent page.

It is this readiness for combat which is taken advantage of to find the nest and eggs, and all that is necessary is to tie a string to your hat, and wave it round your head, when, if a female be sitting in the neighbourhood, the male will instantly come down upon you; and by watching his return the nest may be detected.

Many really absurd statements have been made as to the means by which these birds are obtained for our cabinets. It is most frequently asserted that they are shot with water or with sand. Now, so far as I am aware, these devices are never resorted to, but they are usually procured in the ordinary way, with numbers ten and eleven shot, those being the sizes best suited for the purpose. If smaller shot be used, the plumage is very frequently so cut and damaged that the specimen is rendered of little or no value. By far the greater number fall to the clay ball of the blowpipe, which the Indians, and in some instances even Europeans use with perfect certainty of aim. My friend Professor Jameson has a son who appears to be a proficient in this mode of obtaining Humming-Birds, as I know that many of the specimens he has sent me have been thus procured.

In Brazil very fine nets are employed for this purpose, but how this engine is employed I am unable to state; unfortunately for me many specimens of the fine species *Cometes sparganurus* in my possession have been obtained by means of birdlime, and this is evidently the way in which these birds are captured in the neighbourhood of Chuquisaca.

That the Humming-Bird is not altogether denied the power of song we learn from the notices respecting its vocalization by various authors; but as this is a point upon which I cannot speak from personal observation, I shall take the liberty of quoting from those who have written on the subject. To begin with the remarks of my friend Mr. W. C. L. Martin:—

"It is not to the most beautiful birds that the voice of melody is given. The Mocking-Bird, the Nightingale, and the Thrush, are but plainly attired; and it would appear that if Nature be lavish in one respect, she is parsimonious in another. On the Humming-Birds she has bestowed the gift of beauty—she has created them winged gems—she has chased their plumage with burnished metals or overspread it with laminæ of topaz and emerald—she has strained, so to speak, at every variety of effect—she has revelled in an infinitude of modifications, whether we look at the hues or the development of the feathering. We can scarcely, then, expect that, to such an external perfection, the gift of song will be also added; and, indeed, when we reflect upon the structure of the tongue, of the os hyoides which supports its base, and of the mechanism by which it is rendered capable of protrusion, remembering that the os hyoides is connected with the larynx, we cannot in reason suppose that these birds can be eminent as songsters. Nevertheless it would appear that some species at least utter, while perched, a sort of querulous warble.

"The ordinary cry of the Humming-Birds is sharp and shrill, generally uttered on the wing, and frequently reiterated by the males during their combats with each other. It is principally, says Lesson, in passing from one place to another, that their cry, which he likens to the syllables *tère-tère*, articulated with more or less force, is excited. Most frequently, he says, they are completely dumb; and he adds that he has passed whole hours in observing them in the forests of Brazil without having heard the slightest sound proceed from their throats."

Mr. Gosse, in his 'Birds of Jamaica,' speaking of a species which he calls the Vervain Humming-Bird (the *Mellisuga minima* of this work), says, "The present is the only Humming-Bird that I am acquainted with that has a real song. Soon after sunrise, in the spring months, it is fond of sitting on the topmost branch of a mango or orange-tree, where it warbles in a very weak, but very sweet tone, a continuous melody for ten minutes at a time; it has little variety. The others only utter a pertinacious chirping."

It will be expected that some remarks should now be made with regard to the luminous character of certain parts of the plumage of these charming birds—a point which has engaged the attention of many naturalists and physiologists, but of which I believe no very satisfactory solution has yet been attained. "A few days since," says Mr. Martin, "we were examining a Humming-Bird, the gorget of which was an intense emerald-green, but on changing the light (that is altering its angle of incidence) the emerald was changed into velvet-black. Audebert considered this changeableness to be due to the organization of the feathers, and to the manner in which the luminous rays are reflected on falling upon them; and of this we think there can be little doubt, for each feather, when minutely inspected, exhibits myriads of little facets so disposed as to present so many angles to the incidence of light, which will be diversely reflected according to the position of the feather, and in some positions not reflected in any sensible degree, and thus emerald may become a velvet-black.

"Lesson supposes that the brilliant hues of the plumage of the Humming-Birds are derived from some elements contained in the blood, and elaborated by the circulation—a theory we do not quite understand, inasmuch as colour is the result of the reflection of some rays and the absorption of others, caused by the arrangement of the molecules of any given body. He adds, however, that the texture of the plumes plays the principal part, in consequence of the manner in which the rays of light traverse them, or are reflected by the innumerable facets which a prodigious quantity of barbules or fibres present. All the scaly feathers, he observes, which simulate velvet, the emerald, or the ruby, and which we see on the head and throat of the *Epimachi* (as the Grand Promerops of New Guinea), the Paradise-Birds, and the Humming-Birds, resemble each other in the uniformity of their formation; all are composed of cylindrical barbules, bordered with other analogous regular barbules, which, in their turn, support other small ones, and all of them are hollowed in the centre with a deep furrow, so that when the light, as Audebert first remarked, glides in a vertical direction over the scaly feathers, the result is that all the luminous rays are absorbed in traversing them, and the perception of black is produced. But it is no longer the same when the light is reflected from these feathers, each of which performs the office of a reflector; then it is that the aspect of the emerald, the ruby, &c. varying with the utmost diversity under the incidences of the rays which strike them, is given out by the molecular arrangement of the barbules. It is thus that the gorget of many species takes all the hues of green, and then the brightest and most uniformly golden tints down to intense velvet-black, or, on the contrary, that of ruby, which darts forth pencils of light, or passes from reddish orange to a crimsoned red-black.

"It is thus, we think, that the everchanging hues of the gorgets of the Humming-Birds from black to emerald, ruby, crimson, or flame colour are to be explained."

In a note just received from Dr. Davy, dated Ambleside, June 10, 1861, that gentleman says:—"I have examined with the microscope the feathers of the Humming-Bird, *Aglæactis cupripennis*, you entrusted to me, which is so remarkable for its rich colours as seen in one direction, and only one. The result is merely the following—viz., that those feathers in which this peculiarity is most strongly marked are membranous, terminating in pointed filaments, set on obliquely, so that looking from the head each feather is only partially seen. This result, I apprehend, will help very little to account for the peculiarity in question. Its explanation must be sought (must it not?) in the higher optics."

"As to the question you ask me about the beautiful play of colours in the Humming-Birds," says Dr. Stevelly, "I have never studied the subject, and I should greatly fear to say anything about it, particularly if what I said were to be looked on as of any authority.

"There are two optical principles only which I can see to be any way concerned in such an effect. One is the cause of the play of colours in mother-of-pearl, and which Brewster proved to arise from very fine striated rulings, the distance between the parallel lines not being greater than from the 10,000th to the 100,000th of an inch. Barton, of Birmingham, imitated this by ruling very fine parallel lines on steel dies, and then impressing these on buttons, which showed very beautiful colours when exposed to strong light. The other optical principle, which I think, however, to be the most likely to produce the effect in the case of feathers, is the influence of thin plates. If you know Mr. Gassiot (one of your leading Royal Institution savants) get him to show you some of his copper-plates, on which by an electrotype process he has had very thin films of lead deposited; and I think you will see colours fully as beautiful, though not as varied or as variable in different aspects as those of the Humming-Bird."

It may not be out of place now to give a few extracts from the works of those authors who have written on the *Trochilidæ* in general or on some particular species. A perusal of these will tend to confirm much that I have said; and it is but fair that the writings of those who have wielded the pen in elucidation of the history, habits, and manners of these lovely birds should be duly recognized.

It is fortunate for the science of Ornithology that so many persons gifted with the power of expressing their ideas in elegant and poetical language should have bestowed a large share of their attention upon the Humming-Bird. The writings of Buffon, Wilson, Waterton, Audubon, Gosse, and others, treating exclusively on natural history, are not, perhaps, so generally known as they ought to be; the extracts from these authors will therefore, I doubt not, be found of interest.

"Of all animated beings," says Buffon, "this is the most elegant in form and the most brilliant in colour. The stones and metals polished by art are not comparable to this gem of nature: she has placed it in the order of birds, but among the tiniest of the race—*maxime miranda in minimis*; she has loaded it with all the gifts of which she has only given other birds a share. Agility, rapidity, nimbleness, grace, and rich attire, all belong to this little favourite. The emerald, the ruby, and the topaz, glitter in its garb, which is never soiled with the dust of earth; for, leading an aërial life, it rarely touches the turf even for an instant. Always in the air flying from flower to flower, it shares their freshness and their splendour, lives on their nectar, and only inhabits those climates in which they are unceasingly renewed. The Humming-Bird seems to follow the sun, to advance, to retire with him, and to fly on the wings of the wind in pursuit of an eternal spring."

"Nature in every department of her works," says Wilson, "seems to delight in variety; and the present subject is almost as singular for its minuteness, beauty, want of song, and manner of feeding, as the preceding (the Mocking-Bird) is for unrivalled excellence of notes and plainness of plumage. This is one of the few birds that are universally beloved; and amidst the sweet dewy serenity of a summer's morning, his appearance among the arbours of honey-suckles and beds of flowers is truly interesting.

> "When morning dawns, and the blest sun again
> Lifts his red glories from the eastern main,
> Then through our woodbines, wet with glittering dews,
> The flower-fed Humming-Bird his round pursues;
> Sips with inserted tube the honied blooms,
> And chirps his gratitude as round he roams;
> While richest roses, though in crimson drest,
> Shrink from the splendour of his gorgeous breast.
> What heavenly tints in mingling radiance fly!
> Each rapid movement gives a different dye;
> Like scales of burnished gold they dazzling show—
> Now sink to shade, now like a furnace glow!"

"Where is the person," says Audubon, speaking of the *Trochilus colubris*, "who, on seeing this lovely little creature moving on humming winglets through the air, suspended as if by magic in it, flitting from one flower to another, with motions as graceful as they are light and airy, pursuing its course and yielding new delights wherever it is seen—where is the person, I ask, who, on observing this glittering fragment of the rainbow, would not pause, admire, and turn his mind with reverence towards the Almighty Creator, the wonders of whose hand we at every step discover, and of whose sublime conceptions we everywhere observe the manifestations in his admirable system of creation? There breathes not such a person; so kindly have we all been blessed with that intuitive and noble feeling—admiration.

"I wish it were in my power to impart to you, kind reader, the pleasures which I have felt while watching the movements and viewing the manifestations of feelings displayed by a single pair of these most favourite little creatures, when engaged in the demonstration of their love for each other;—how the male swells his plumage and throat, and, dancing on the wing, whirls around the delicate female; how quickly he dives towards a flower, and returns with a loaded bill, which he offers to her to whom alone he desires to be united; how full of ecstacy he seems to be when his caresses are kindly received; how his little wings fan her as they fan the flowers, and he transfers to her bill the insect and the honey which he has procured with a view to please her; how these attentions are received with apparent satisfaction; how, soon after, the blissful compact is sealed; how, then, the courage and care of the male is redoubled; how he even dares to give chase to the tyrant Flycatcher, hurries the Blue-Bird and the Martin to their boxes; and how, on sounding pinions, he joyously returns to the side of his lovely mate. Reader, all these proofs of the sincerity, fidelity, and courage with which the male assures his mate of the care he will take of her while sitting on her nest, may be seen, have been seen, but cannot be pourtrayed or described.

"Could you cast a momentary glance on the nest of the Humming-Bird and see, as I have seen, the newly hatched pair of young, little larger than humble-bees, naked, blind, and so feeble as scarcely to be able to raise their little bill to receive food from the parents; and could you see those parents full of anxiety and fear, passing and repassing within a few inches of your face, alighting on a twig not more than a yard from your body, waiting the result of your unwelcome visit in a state of the utmost despair, you could not fail to be impressed with the interest of the scene. Then how pleasing it is, on your leaving the spot, to see the returning hope of the parents when, after examining the nest, they find their nestlings untouched! These are the scenes best fitted to enable us to partake of sorrow and joy, and to determine every one who views them to make it his study to contribute to the happiness of others, and to refrain from wantonly or maliciously giving them pain.

"A person standing in a garden by the side of a common Althæa in bloom, will be surprised to hear the humming of their wings, and then see the birds themselves within a few feet of him, as he will be astonished at the rapidity with which the little creatures rise into the air, and are out of sight and hearing the next moment.

"No bird seems to resist their attacks; but they are sometimes chased by the larger kinds of humble-bees, of which they seldom take the least notice, as their superiority of flight is sufficient to enable them to leave those slow-moving insects far behind in the short space of a minute.

"If comparison might enable you to form some tolerably accurate idea of their peculiar mode of flight, and their appearance when on the wing, I should say that, were both objects of the same colour, a large *Sphinx* or moth when moving from one flower to another, and in a direct line, comes nearer the Humming-Bird in aspect than any other object with which I am acquainted."—*Audubon, Ornithological Biography*, vol. i. p. 248, &c. For the other portions of Wilson's and Audubon's very interesting observations, I must refer my readers to my account of *Trochilus colubris*.

"Though least in size," remarks Mr. Waterton, "the glittering mantle of the Humming-Bird entitles it to the first place in the list of the birds of the New World. It may truly be called the Bird of Paradise; and had it existed in the Old World it would have claimed the title, instead of the bird which has now the honour to bear it. See it darting through the air almost as quick as thought!—now it is within a yard of your face!—in an instant it is gone!—now it flutters from flower to flower to sip the silver dew—it is now a ruby —now a topaz—now an emerald—now all burnished gold! It would be arrogant to pretend to describe this winged gem of nature after Buffon's elegant description of it.

"Cayenne and Demerara produce the same Humming-Birds. Perhaps you would wish to know something of their haunts. Chiefly in the months of July and August, the tree called Bois Immortel, very common in Demerara, bears abundance of red blossom, which stays on the tree for some weeks; then it is that most of the species of Humming-Birds are very plentiful. The wild red sage (*Salvia splendens*) is also their favourite shrub; and they buzz like bees round the blossom of the Wallaba-tree; indeed there is scarce a flower in the interior, or on the sea-coast, but what receives frequent visits from one or other of the species.

"On entering the forests of the rising land in the interior, the blue and green, the smallest brown, no bigger than the humble-bee, with two long feathers in the tail, and the little forked-tail purple-throated Humming-Birds glitter before you in everchanging attitudes.

"As you advance towards the mountains of Demerara, other species of Humming-Birds present themselves before you. It seems to be an erroneous opinion that the Humming-Bird lives entirely on honey-dew. Almost every flower of the tropical climate contains insects of one kind or other; now the Humming-Bird is most busy about the flowers an hour or two after sun-rise, and after a shower of rain; and it is just at this time that the insects come out to the edge of the flower in order that the sun's rays may dry the nocturnal dew and rain which they have received. On opening the stomachs of the Humming-Bird dead insects are almost always found there."

"The Humming-Birds in Jamaica," says Lady Emmeline Stuart Wortley in her Travels, "are lovely little creatures, and most wonderfully tame and fearless of the approach of man. One of these charming feathered jewels had built its delicate nest close to one of the walks of the garden belonging to the house where we were staying. The branch, indeed, of the beautiful little shrub in which this fairy nest was suspended almost intruded into the walk; and every time we sauntered by there was much danger of sweeping against this projecting branch with its precious charge, and doing it some injury, as very little would have demolished the exquisite fabric; in process of time, two lovely little pear-like eggs had appeared; and while we were there we had the great pleasure of seeing the minute living gems themselves appear, looking like two very small bees. The mother-bird allowed us to look closely at her in the nest, and to inspect her little nurslings, when she was flying about near, without appearing in the least degree disconcerted or alarmed. I never saw so tame or so bold a little pet. But she did not allow the same liberties to be taken by everybody unchecked. One day, as Sir C—— was walking in the pretty path beside which the fragile nest was delicately suspended amid sheltering leaves, he paused, in order to look at its Lilliputian inhabitants. While thus engaged, he felt suddenly a sharp light rapping on the crown of his hat, which considerably surprised him. He looked round to ascertain from whence the singular and unexpected attack proceeded: but nothing was to be seen. Almost thinking he must have been mistaken, he continued his survey; when a much sharper and louder rat-tat-tat-tat-tat seemed to demand his immediate attention, and a little to jeopardize the perfect integrity and preservation of the fabric in question. Again he looked round, far from pleased at such extraordinary impertinence; when what should he see but the beautiful delicate Humming-Bird, with ruffled feathers and fiery eyes, who seemed by no means inclined to let him off without a further infliction of sharp taps and admonitory raps from her fairy beak. She looked like a little fury in miniature—a winged Xantippe. Those pointed attentions apprised him that his company was not desired or acceptable; and, much amused at the excessive boldness of the dauntless little owner of the exquisite nest he had been contemplating, Sir C—— moved off, anxious not to disturb or irritate further this valiant minute mother, who displayed such intrepidity and cool determination. As to V—— and me, the darling little pet did not mind us in the least; she allowed us to watch her to our hearts' content during the uninterrupted progress of all her little household and domestic arrangements, and rather appeared to like our society than not, and to have the air of saying, 'Do you think I manage it well, eh?'"

"I cannot quit the subject," says the Reverend Lansdown Guilding, "without speaking of the delight that was afforded me, in Jamaica, by seeing Humming-Birds feeding on honey in the florets of the great Aloe (*Agave Americana*, Linn.) On the side of a hill upon Sutton's Estate (the property of Henry Dawkins, Esq.) were a considerable number of aloe plants, of which about a dozen were in full blossom. They were spread over a space of about twenty yards square. The spikes bearing bunches of flowers in a thyrsus, were from twelve to fifteen feet high; on each spike were many hundred flowers of a bright yellow colour, each floret of a tubular shape and containing a good-sized drop of honey. Such an assemblage of floral splendour was in itself most magnificent and striking; but it may be imagined how much the interest caused by this beautiful exhibition was increased by vast numbers of Humming-Birds, of various species fluttering at the opening of the flowers, and dipping their bills first into one floret and then into another,—the sun, as usual, shining bright upon their varied and beautiful plumage. The long-tailed or Bird-of-Paradise Humming-Bird was particularly striking, its long feathers waving as it darted from one flower to another. I was so much delighted with this sight that I visited the spot again in the afternoon, after a very long and fatiguing day's ride, accompanied by my wife, on horseback, when we enjoyed the scene before us for more than half-an-hour."

"The pugnacity of the Humming-Birds," remarks Mr. Gosse, "has been often spoken of; two of one species can rarely suck flowers from the same bush without a rencontre. I once witnessed a combat between two, which was prosecuted with much pertinacity, and protracted to an unusual length. It was in the month of April, when

I was spending a few days at Phœnix Park, near Savannah la Mar, the residence of my kind friend Aaron Deleon, Esq. In the garden were two trees, of the kind called Malay Apple (*Eugenia Malaccensis*), one of which was but a yard or two from my window. The genial influence of the spring rains had covered them with a profusion of beautiful blossoms, each consisting of a multitude of crimson stamens, with very minute petals, like bunches of crimson tassels; but the leaf-buds were only beginning to open. A Humming-Bird had every day and all day long been paying his devoirs to these charming blossoms. On the morning to which I allude another came, and the manœuvres of these two tiny creatures became very interesting. They chased each other through the labyrinths of twigs and flowers till, an opportunity occurring, the one would dart with seeming fury upon the other, and then, with a loud rustling of their wings, they would twirl together, round and round, till they nearly came to the earth. It was some time before I could see, with any distinctness, what took place in these tussles; their twirlings were so rapid as to baffle all attempts at discrimination. At length an encounter took place pretty close to me, and I perceived that the beak of the one grasped the beak of the other, and thus fastened both whirled round and round in their perpendicular descent, the point of contact being the centre of the gyrations, till, when another second would have brought them both on the ground, they separated, and the one chased the other for about a hundred yards and then returned in triumph to the tree, where, perched on a lofty twig, he chirped monotonously and pertinaciously for some time—I could not help thinking in defiance. In a few minutes, however, the banished one returned and began chirping no less provokingly, which soon brought on another chase and another tussle. I am persuaded that these were hostile encounters; for one seemed evidently afraid of the other, fleeing when the other pursued, though his indomitable spirit would prompt the chirp of defiance; and when resting after a battle, I noticed that this one held his beak open, as if panting. Sometimes they would suspend hostilities to suck a few blossoms, but mutual proximity was sure to bring them on again, with the same result. In their tortuous and rapid evolutions the light from their ruby necks would occasionally flash in the sun with gem-like radiance; and as they now and then hovered motionless, the broadly expanded tail, the outer feathers of which are crimson-purple, but when intercepting the sun's rays transmit orange-coloured light, added much to their beauty. A little Banana Quit (*Certhiola flaveola*), that was peeping among the blossoms in his own quiet way, seemed now and then to look with surprise on the combatants; but when the one had driven his rival to a longer distance than usual, the victor set upon the unoffending Quit, who soon yielded the point, and retired, humbly enough, to a neighbouring tree. The war (for it was a thorough campaign, a regular succession of battles) lasted fully an hour, and then I was called away from the post of observation. Both of the Humming-Birds appeared to be males."

"All the Humming-Birds have more or less the habit, when in

flight, of pausing in the air and throwing the body and tail into rapid and odd contortions. This is most observable in the Polytmus, from the effect that such motions have on the long feathers of the tail. That the object of these quick turns is the capture of insects I am sure, having watched one thus engaged pretty close to me. I observed it carefully, and distinctly saw the minute flies in the air which it pursued and caught, and heard repeatedly the snapping of the beak. My presence scarcely disturbed it, if at all."

In some notes on the 'Habits of the Humming-Birds of the Amazon,' kindly furnished me by Mr. Wallace, that gentleman says—

"The great number of species that frequent flowers, do so, I am convinced, for the small insects found there, and not for the nectar. In dozens, and perhaps hundreds, of common flower-frequenting species which I have examined, the crop, stomach, and intestines have been entirely filled with minute beetles, bees, ants, and spiders, which abound in most flowers in South America. Very rarely, indeed, have I found a trace of honey or of any liquid in the crop or stomach. The flowers they most frequent are the various species of *Inga*, and the papilionaceous flowers of many large forest-trees. I have never seen them at the Bignonias or any flowers but those which grow in large masses covering a whole tree or shrub; as they visit perhaps a hundred flowers in a minute and never stop at a single one. The little Emerald Hummer I have seen in gardens and at the common orange *Asclepias*, which often covers large spaces of waste ground in the tropics. But there are many, such as *Phaëthornis Eremita*, and some larger allied species, which I have never seen at flowers. These inhabit the gloomy forest-shades, where they dart about among the foliage, and I have distinctly observed them visit in rapid succession every leaf on a branch, balancing themselves vertically in the air, passing their beak closely over the under surface of each leaf, and thus capturing, no doubt, any small insects that may be upon them. While doing this the two long feathers of their tail have a vibrating motion, serving apparently as a rudder to assist them in performing the delicate operation. I have seen others searching up and down stems and dead sticks in the same manner, every now and then picking off something exactly as a Bush-strike, or a Tree-creeper does, with this exception that the Humming-Bird is constantly on the wing. They also capture insects in the true fissirostral manner. How often may they be seen perched on the dead twig of a lofty tree—the same station that is chosen by the tyrant Flycatchers and the Jacamars, and from which, like those birds, they dart off a short distance and, after a few whirls and balancings, return to the identical twig they had left. In the evening too, just after sunset, when the Goat-suckers are beginning their search after insects over the rivers, I have seen Humming-Birds come out of the forest and remain a long time on the wing, now stationary, now darting about with the greatest rapidity, imitating in a limited space the varied evolutions of their companions the Goat-suckers, and evidently for the same end and purpose.

"Many naturalists have noticed this habit of feeding on insects, but have generally considered it as the exception, whereas I am inclined to think it is the rule. The frequenting of flowers seems to me only one of the many ways by which they are enabled to procure their insect-food."

"Wilson, Audubon, Mr. Gosse, and several others gifted with the 'pen of a ready writer,'" says Mr. Alfred Newton, "have so fully described, as far as words will admit, the habits of different members of the family *Trochilidæ*, that it is unnecessary to say much upon this score. Their appearance is so entirely unlike that of any other birds, that it is hopeless to attempt in any way to bring a just conception of it to the ideas of those who have not crossed the Atlantic; and even the comparison so often made between them and the *Sphingidæ*, though doubtless in the main true, is much to the advantage of the latter. One is admiring the clustering stars of a scarlet *Cordia*, the snowy cornucopias of a *Portlandia*, or some other brilliant and beautiful flower, when between the blossom and one's eye suddenly appears a small dark object, suspended as it were between four short black threads meeting each other in a cross. For an instant it shows in front of the flower; an instant more, it steadies itself, and one perceives the space between each pair of threads occupied by a grey film; again another instant, and emitting a momentary flash of emerald and sapphire light, it is vanishing, lessening in the distance, as it shoots away, to a speck that the eye cannot take note of,—and all this so rapidly that the word on one's lips is still unspoken, scarcely the thought in one's mind changed. It was a bold man or an ignorant one who first ventured to depict Humming-Birds flying; but it cannot be denied that representations of them in that attitude are often of special use to the ornithologist. The peculiar action of one, and probably many or all other species of the family, is such, that at times, in flying, it makes the wings almost meet, both in front and behind, at each vibration. Thus when a bird chances to enter a room, it will generally go buzzing along the cornice; standing beneath where it is, one will find that the axis of the body is vertical, and each wing is describing a nearly perfect semicircle. As might be expected, the pectoral muscles are very large; indeed the sternum of this bird is a good deal bigger than that of the common Chimney Swallow (*Hirundo rustica*, L.). But the extraordinary rapidity with which the vibrations are effected seems to be chiefly caused by these powerful muscles acting on the very short wing-bones, which are not half the length of the same parts in the Swallow; and accordingly great as this alar action is, and in spite of the contrary opinion entertained by Mr. Gosse (Nat. Sojourn in Jamaica, 240), it is yet sometimes wanting in power, owing doubtless to the disadvantageous leverage thus obtained; and the old authors must be credited who speak of cobwebs catching Humming-Birds.

"On the 3rd of May, 1857, a bird of this species" (*Eulampis chlorolæmus*, Gould) "flew into the room where I was sitting, and, after fluttering for some minutes against the ceiling, came in contact with a deserted spider's web, in which it got entangled, and remained suspended and perfectly helpless for more than a minute, when by a violent effort it freed itself. I soon after caught it, still

D 2

having fragments of the web on its head, neck, and wings; and I feel pretty sure that had this web been inhabited and in good repair, instead of being deserted and dilapidated, the bird would never have escaped."—A. N.

In his 'Notes on the Humming-Birds of Guatemala,' Mr. Salvin says, " During the months of August and September the localities of the various species of Humming-Birds are usually as follows:— Among the trees on the south-eastern side of the lake" of Dueñas "are *Amazilia Devillei, Thaumastura henicura* (mostly females), *Campylopterus rufus, Heliomaster longirostris, Chlorostilbon Osberti* (in small number), *Cyanomyia cyanocephala*, and *Trochilus colubris*.

"On the hill-side to the south-westward of the lake are great numbers of *Campylopterus rufus*, and among the willows close to the water the males of *Thaumastura henicura* congregate. About the Convolvulus-trees in the llano at the foot of the volcano are found *Eugenes fulgens, Amazilia Devillei, Thaumastura henicura* (in small numbers), *Trochilus colubris* (very commonly towards the end of September).

"Entering the first barranco that opens out into the plain, we meet with *Campylopterus rufus, Myiabeillia typica, Heliopædica melanotis*, and a little higher up, *Petasophora thalassina* and *Delattria viridipallens*. Of course, occasionally a species is found not in its place as here indicated; for instance, I have seen in the first locality a single specimen (the only female I have met with) of *Eugenes fulgens*, and another high in the volcano. I have also seen a single *Petasophora thalassina* out on the llano. These localities must therefore be taken as only generally indicating the distribution of the species found about Dueñas."—*Ibis*, vol. ii. p. 263.

At the moment of printing these pages, I have received a very interesting letter from my friend the Hon. G. W. Allan, of Moss Park, Toronto, in which the following passage occurs respecting the *Trochilus colubris*:—

"I wish you could have been with us last summer, you would have had an opportunity of watching your favourite Humming-Birds to your heart's content. I do not in the least exaggerate when I say that, during the time the horse-chestnuts were in flower, there were hundreds of these little tiny creatures about my grounds. While sitting in my library I could hear their little sharp, querulous note, as the males fought like so many little bantam cocks with each other. On one large horse-chestnut tree, just at the corner of the house, they swarmed about the foliage like so many bees; and as the top branches of the tree were close to my bed-room windows, every now and then one bird, more bold than the rest, would dart into the open window and perch upon the wardrobe or the top of the bed-post."

It will be expected that, in a monograph of a group of birds which have attracted so much notice, some account should be given of their internal structure, and as our well-known bird-anatomist, T. C. Eyton, Esq., who has paid much attention to the subject, has given a very clear description of that of the largest species of the family—the

Patagona gigas—in Mr. Darwin's 'Zoology of the Voyage of H. M. S. Beagle,' I have much pleasure in transferring it to my pages:—

"Tongue bifid, each division pointed; hyoids very long, in their position resembling those in the *Picidæ* (Woodpeckers); trachea of uniform diameter, destitute of muscles of voice; bronchia very long; œsophagus funnel-shaped, slightly contracted on approaching the proventriculus, which is small and scarcely perceptible; gizzard small, moderately muscular, the inner coat slightly hardened, and filled with the remains of insects; intestine largest near the gizzard; I could not perceive a vestige of cæca. Length of the œsophagus, including the proventriculus, 1¾ inch; of the intestinal canal 3½; length of the gizzard ½, breadth ¼.

"Sternum with the keel very deep, its edge rounded and projecting anteriorly; posterior margin rounded, and destitute of indentation or fissure; the ridges to which the pectoral muscles have their attachment large and prominent, the horizontal portion much narrowed anteriorly, consequently the junctions of the coracoids are very near together.

"Pelvis short, very broad; os pubis long, curved upwards at the extremities, projecting far downwards, and posteriorly beyond the termination of the caudal vertebræ; the ischiatic foramen small and linear; femora placed far backwards; coracoids short, very strong, their extremities much diverging; os furcatum short, slightly arched near the extremities of the rami, which are far apart, furnished with only a small process on its approach to the sternum; scapula flattened, long, broadest near the extremity; humerus, radius, and ulna short, the metacarpal bones longer than either; the former furnished with ridges much elevated for the attachment of the pectoral muscles; caudal and dorsal vertebræ with the transverse processes long and expanded; cranium of moderate strength, the occipital portion indented with two furrows, which pass over the vertex, and in which the hyoids lie; orbits large, divided by a complete bony septum; the lacrymal bones large, causing an expansion of the bill near the nostrils.

"Number of cervical vertebræ 10, dorsal 6, sacral 9, caudal 5; total 30.

"Number of true ribs 5, false 4; total 9."

Dr. Davy states that the blood-corpuscles of a recently killed Humming-Bird, examined by him in Barbadoes, "were beautifully definite, regular and uniform. The disk very thin, perfectly flat; the nucleus slightly raised, and the two corresponding in outline. The corpuscles 1–2666th by 1–4000th of an inch, the long diameter of the nucleus very nearly 1–4000th. The blood was small in quantity, as I apprehend is the blood of birds generally, but not deficient in red corpuscles. I have found its temperature to be about 105 degrees."

I have found it impossible to divide the Humming-Birds into more than two subfamilies—*Phaëthornithinæ* and *Trochilinæ*—for I find no such well-marked divisions among them as will enable me so to do; neither can I arrange them in anything like a continuous series; so many gaps occur here and there, that one is almost led to the belief that many forms have either died out or have not yet been discovered; consequently I am unable to commence with any one genus and arrange the remainder in accordance with their affinity. Whenever I have observed an apparent relationship between two or more genera, they have been placed in contiguity, and the species which appear to be allied to each other are arranged in continuous succession. I do not consider one species more typical than another; all are equally and beautifully adapted for the purposes they are intended to perform.

The following Synopsis will be found to contain a general view of the subject; and as it also comprises the additional information I have been able to obtain during the progress of the work, should always be consulted.

I shall now give the general characters by which the *Trochilidæ* are distinguished:—

Body small; sternum very deep; bill subulate, and generally longer than the head, straight, arched, or upcurved; tongue composed of two lengthened cylindrical united tubes, capable of great protrusion, and bifid at the tip; nostrils basal, linear, and covered by an operculum; wings lengthened, pointed, the first of which is the longest, except in the genus *Aithurus*, where it is the second; primaries ten in number; tarsi and feet very diminutive; tail consisting of ten feathers. The entire structure adapted for aërial progression.

Subfamily I. PHAËTHORNITHINÆ.

I commence my first volume with that well-marked section of the family comprising the genera *Grypus, Eutoxeres, Phaëthornis*, and their allies. The members of all these genera are remarkable for being destitute of metallic brilliancy, and, as their trivial name of "hermits" implies, for affecting dark and gloomy situations. They constitute, perhaps, the only group of the great family of Humming-Birds which frequent the interior of the forests, and there obtain their insect food, some from the underside of the leaves of the great trees, while others assiduously explore their stems in search of such lurking insects as may be concealed in the crevices of the bark. It has been said that spiders constitute the food of many species of this group; and I believe that such is the case, for we find the bills admirably adapted for their capture, particularly those of *Grypus* and *Eutoxeres*. To individualize by name any particular country in South America in which these birds are found is unnecessary, for they are generally distributed over its temperate and hotter portions; but they are not to be met with either very far north or very far south of the equator; that is to say, their range is bounded northwardly by Southern Mexico, and southwardly by Bolivia. Within these limits, the high and the low lands are alike tenanted by them; but it is in the equatorial region that they are the most numerous, and where all, or nearly all, the genera have representatives. In the colouring of their plumage the sexes are generally alike. As a whole, they form a well-marked division distinguished by their own especial peculiarities of form and style of plumage.

Genus GRYPUS, *Spix.*

This form, which comprises two species, both natives of Brazil, is remarkable for the bill of the male being different in structure from that of the female.

1. GRYPUS NÆVIUS . . . . Vol. I. Pl. I.

Trochilus nævius, Dumont, Temm., Vieill., Drap., Burm.
——— *squamosus*, Licht.
Grypus ruficollis, Spix.
Mellisuga nævia, Steph.
Ramphodon maculatum, Less.
——— *nævius*, Less., Jard.
Grypus nævius, Gray & Mitch., Bonap.
*Phæthornis nævius, Jard. Nat. Lib. Humming-Birds, vol. ii. p. 152.
*Ramphodon nævius, Reichenb. Aufz. der Col. p. 15; Id. Troch. Enum. p. 12; Cab. et Hein. Mus. Hein. Theil iii. p. 3.

Habitat. South-eastern Brazil.

2. GRYPUS SPIXI, *Gould* . . . . Vol. I. Pl. II.

*Ramphodon chrysurus, Reichenb. Aufz. der Col. p. 15; Id. Troch. Enum. p. 12?

Habitat. Brazil.

The law of adaptation is perhaps equally carried out in every one of the multiplied forms, not only of ornithology, but of every other department of nature's works, each being constructed for some given purpose contributing to the well-being of the animal; in some instances, however, particular developments are more striking and singular than in others. The form to which the generic name of *Eutoxeres* has been given is a case in point. Of this remarkable genus two species are known, both of which are natives of the Andes of Ecuador, New Granada, and Veragua. It would be most interesting to become acquainted with their peculiar modes of life, and to ascertain for what end their singularly curved bills were designed. Some persons affirm that it is for the purpose of probing the scaly covering of the upright stems of certain trees, and others for the exploration of peculiar cup-shaped flowers, such as that of the orchid which I have figured in the plate of *Eutoxeres Aquila.* Whatever may be the design, future research must determine it; all that we at present know is, that this form does exist, and that there is none other which approaches to it. In size the two species are very similar, but there are good and plain specific characters by which they may be distinguished, and which will, I trust, be sufficiently apparent on reference to the plates in which the birds are represented.

Genus EUTOXERES, *Reichenb.*

The oldest-known species of this form is the—

3. EUTOXERES AQUILA . . . . Vol. I. Pl. III.

Trochilus Aquila, Lodd., Boure.
Politmus Aquila, Gray & Mitch.
Glaucis Aquila, Bonap.
* *Eutoxeres Aquila*, Reichenb. Aufz. der Col. p. 15; Id. Troch. Enum. p. 12; Cab. et Hein. Mus. Hein. Theil iii. p. 3, note.
* *Myiaëtina aquila*, Bonap. Rev. et Mag. de Zool. 1854, p. 249.

Habitat. Costa Rica, New Granada, and Ecuador.

The following notes respecting this species by Dr. J. King Merritt will be read with interest. They are extracted from the 6th volume of the 'Annals of the Lyceum of Natural History of New York,' p. 139:—

"It was, as near as I can recollect, during the month of September 1852 that I saw for the first time and obtained a specimen of this (to me) curious and novel bird. I was at that time stationed in the mountainous district of Belen, province of Veragua, New Granada.

"My attention at that particular period was directed towards the collection of specimens of the Humming-Bird family. One day, while out hunting a short distance from the camp, I was startled by the swift approach of a small object through the close thicket, which darted like a rifle bullet past me, with a loud hum and buzzing of wings. Indeed, it was this great noise that accompanied its flight that especially attracted my attention as something uncommon.

"The bird continued its flight but a short distance beyond the spot where I stood, when it suddenly stopped in its rapid course directly in front of a flower. There for a moment poising itself in this position, it darted upon the flower in a peculiar manner; in fact, the movements which now followed were exceedingly curious. Instead of inserting its beak into the calyx by advancing in a direct line towards the flower, as customary with this class of birds, this one performed a curvilinear movement, at first stooping forward while it introduced its beak into the calyx, and then, when apparently the point of the beak had reached the desired locality in the flower, its body suddenly dropped downwards, so that it seemed as though it was suspended from the flower by the beak. That this was not actually the case, the continued rapid movement of its wings demonstrated beyond a doubt. In this position it remained the ordinary length of time, and then, by performing these movements in the reverse order and direction, it freed itself from the flower, and afterwards proceeded to the adjoining one, when the same operation was repeated as already described.

"The flower from which it fed is somewhat peculiar in form, &c. The plant belongs to the Palm species, and grows in low marshy places, on or near the margins of rivers and mountain streams. It consists of a dozen or more straight stems, each of which terminates above in a broad expanded leaf that somewhat resembles the plantain. These stems all start from a clump at the surface of the ground, but they immediately separate, and slightly diverge from each other. The stems with the leaf grow to the height of six to ten feet, more or less. From one or two of the centre stems a flower-stalk puts forth, which hangs pendent, and to this are attached alternately on either side the flowers, while the space between each corresponds with the attachment of the one on the opposite side of the stalk.

"The flower resembles somewhat in form the Roman helmet inverted, and is attached, as it were, by the point of the crest to the stalk. It is a fleshy mass, and the cavity of the calyx extends in a tortuous manner downwards towards the attachment of the flower to the stalk."

4. EUTOXERES CONDAMINEI . . . . Vol. I. Pl. IV.
Trochilus Condamini, Bourc.
**Eutoxeres Condaminei*, Reichenb. Aufz. der Col. p. 15; Id. Troch. Euum. p. 12; Cab. et Hein. Mus. Hein. Theil iii. p. 9, note.
**Myiaëtina condamini*, Bonap. Rev. et Mag. de Zool. 1854, p. 249.

Habitat. Eastern Ecuador.

For the knowledge of the existence of *E. Condaminei* science is indebted to the researches of M. Bourcier, who brought specimens from Archidona.

Genus GLAUCIS, *Boié.*

This genus comprises at least six species, three of which are very nearly allied. It will be seen, on reference to my account of *G. hirsutus*, that when it was written I was much perplexed with regard to its synonymy, or rather, as to whether the small red-coloured bird, *G. mazeppa*, was or was not identical with it; and although some years have since elapsed, I have not even now been able to arrive at a satisfactory solution of the difficulty. Under these circumstances, I think it will be best to regard the *G. mazeppa* as distinct; and this view of the subject is supported by the fact that I do not find small red-coloured birds accompanying the allied species, *affinis*, Lawr., which is a native of Bogota. I think it likely that all these birds, when fully adult, have the tail shorter and more rounded than during the period of immaturity or at the end of the first year of their existence. The youthful state then is indicated by a more cuneate form of tail, all the feathers of which are pointed and tipped with white; and as the birds advance in age —that is, at each moult—the tail-feathers become more rounded and the white tipping less, until at length it is reduced to a mere fringe, existing in some instances on the middle feathers alone.

The distribution of the species of the genus *Glaucis* extends over the whole of the eastern parts of Brazil, the Guianas, Trinidad, Tobago, Venezuela, the banks of the Amazon, New Granada, and Veragua.

5. GLAUCIS HIRSUTA . . . . Vol. I. Pl. V.

Trochilus hirsutus, Gmel. Vieill., Dumont, Temm., Less., Jard.
Phaethornis hirsutus, Jard. & Selb.
Polytmus Brasiliensis, Briss.
Trochilus Brasiliensis, Lath.
Polytmus hirsutus, Gray & Mitch.
Glaucis hirsutus, Boié, Bonap.
——— *hirsuta*, Reich.
* *Trochilus Dominicus*, Licht. Doubl. p. 12; Cab. Schomb. Reise Guian. tom. iii. p. 708.
* ——— *ferrugineus*, Wied, Beitr. iv. p. 20.
* ——— *superciliosus*, fœm., Less. Hist. Nat. des Col. p. 38, pl. 7; Id. Traité d'Orn. p. 289; Jard. Nat. Lib. Humming-Birds, vol. ii. p. 120, pl. 27; Durm. Th. Bras. tom. ii. p. 324.
* ——— *hirsuta*, Cab. et Hein. Mus. Hein. Theil iii. p. 4.

Habitat. Eastern Brazil, Venezuela, and the Island of Trinidad.

6. GLAUCIS MAZEPPA . . . . Vol. I. Pl. VI.

* *Glaucis Mazeppa*, Less. Troch. p. 18, pl. 9; Jard. Nat. Lib. Humming-Birds, vol. ii. p. 150.
* *Phaëthornis Mazeppa*, Jard. Nat. Lib. Humming-Birds, vol. ii. p. 152.
* *Polytmus mazeppa*, Gray & Mitch. Gen. of Birds, vol. ii. p. 108, *Polytmus*, sp. 32.
* *Glaucis Mazeppa*, Reichenb. Aufz. der Col. p. 15.

Habitat. Cayenne, the Guianas, and the Islands of Trinidad and Tobago.

"This little bird," says Mr. Kirk, "is the most restless of all the Humming-Bird tribe; it can scarcely be said to be seen at rest, but darting right and left, zigzag. At times, when suddenly surprised feeding, uttering a sharp squeak it will dart off and disappear like a meteor; at other times it will seem as if suspended for several seconds by the point of the bill within three feet of a person's face, after which it is sure to disappear like lightning; in these cases it truly assumes an attitude which a stranger might construe into a meditated attack upon his person. I have often been induced to strike at them with my fowling-piece from their proximity."—'*Horæ Zoologicæ*,' by Sir W. Jardine, Bart., in *Ann. and Mag. Nat. Hist.* vol. xi. p. 372.

7. GLAUCIS AFFINIS, *Lawr.* . . . . Vol. I. Pl. VII.

* *Glaucis affinis*, Lawr. in Ann. Lyc. Nat. Hist. New York, vol. vi. p. 261.

Habitat. The high lands of New Granada. Specimens are frequently sent from Bogota.

8. GLAUCIS LANCEOLATUS, *Gould* . . . . . Vol. I. Pl. VIII.
Habitat. Para.

9. GLAUCIS MELANURA, *Gould* . . . . . . Vol. I. Pl. IX.
Habitat. The banks of the Rio Napo and the Ilio Negro.

10. GLAUCIS DOHRNI . . . . Vol. I. Pl. X.
Trochilus Dohrnii, Bourc.
Glaucis Dorhni, Bonap., Reich.
*Glaucis Dohrni, Cab. et Hein. Mus. Hein. Theil iii. p. 4.
Habitat. Southern Brazil.

M. Bourcier has given Ecuador as the locality where his specimen was procured; but my bird was received direct from the district of Espirito Santo in Brazil.

11. GLAUCIS RUCKERI . . . . Vol. I. Pl. XI.
Trochilus Ruckeri, Bourc.
Polytmus Ruckeri, Gray & Mitch.
**Threnetes Ruckeri*, Reichenb. Aufz. der Col. p. 15; Id. Troch. Enum. p. 12.
*Glaucis Ruckeri, Cab. et Hein. Mus. Hein. Theil iii. p. 4.
Habitat. Veragua.

12. GLAUCIS FRASERI, *Gould* . . . . . . Vol. I. Pl. XII.
*Glaucis Ruckeri, Sclat. in Proc. of Zool. Soc. part 28. p. 296.

Mr. Fraser collected at Bababoyo, in Ecuador, specimens of a bird which both Dr. Sclater and myself considered to be identical with the *Glaucis Ruckeri*, but which, on a more minute comparison with specimens from Veragua, I find to be sufficiently different to entitle it to be regarded as distinct; I have therefore named it after its discoverer, as a just tribute to one who has played a good part in the furtherance of science. The *G. Fraseri* differs from *G. Ruckeri* in being rather larger in size, in having a smaller amount of rusty red on the chest, and in having a decidedly grey breast; in other respects the two birds are very similar.

The following is Mr. Fraser's note respecting this species:—

" Found on the edge of the virgin forest; always solitary; generally in dark and lonely places, and very restless. Irides hazel; upper mandible black, lower yellow, with a black tip; legs and feet flesh-colour."

Habitat. Ecuador.

Allied to the last form are the members of the genus *Threnetes*; these birds are not distinguished by any brilliancy of colouring, but two of them are very prettily marked about the throat and chest.

Surinam and the adjacent countries are given as the habitat of

T. leucurus, while the banks of the Rio Napo are known to be the home of the bird I have called *cervinicauda*; and the sombre-plumaged *Antoniæ* is a native of Cayenne and the Guianas. I believe that the females of all three species are clothed like the males.

Genus THRENETES, *Gould*.

(Θρηνητὴς, a mourner.)

Generic characters.

Male.—*Bill* lengthened, arched, and pointed; *wings* moderately long, and rounded at the tip; *tail* short, square, or rounded; *tarsi* partially clothed; *feet* very small; *hind toe* and *nail* short.

13. THRENETES LEUCURUS . . . . Vol. I. Pl. XIII.

Trochilus leucurus, Linn., Gmel., Lath., Less., Vieill., Dumont, Drap.
Polytmus Surinamensis, Briss.
——— *leucurus*, Gray & Mitch.
Glaucis leucurus, Bonap.
Threnetes leucurus, Reichenb. Aufz. der Col. p. 15; Id. Troch. Enum. p. 12.

Habitat. Surinam and British Guiana.

14. THRENETES CERVINICAUDA, *Gould* . . . Vol. I. Pl. XIV.

Threnetes cervinicauda, Gould in Proc. of Zool. Soc. part 22. p. 109.

Habitat. Province of Quijos in Ecuador.

15. THRENETES ANTONIÆ . . . . Vol. I. Pl. XV.

Trochilus Antoniæ, Bourc. et Mula.
Polytmus Antoniæ, Gray & Mitch.
Lampornis Antoniæ, Bonap.
Aphantochroa Antoniae, Reichenb. Aufz. der Col. p. 15; Id. Troch. Enum. p. 12.

Habitat. Cayenne and the Guianas.

I now enter upon the genus *Phaëthornis*, the members of which are so widely dispersed, that the remark I made on the extended distribution of the entire group is almost applicable to this section of it. In the body of this work I have figured nearly thirty species under this generic appellation, including therein the smaller kinds to which Bonaparte gave the name of *Pygmornis*, a term I shall now adopt for these little birds: but a further subdivision of the group I cannot for a moment entertain; the separation of the *P. Bourcieri* into a distinct genus, for which the term *Ametrornis* has been proposed by Dr. Reichenbach, and of the *P. Guyi* under that of *Toxotruches* by Dr. Cabanis, being, in my opinion, quite unnecessary.

Genus PHAËTHORNIS, *Swains.*

16. PHAËTHORNIS EURYNOME . . . . . . Vol. I. Pl. XVI.

Trochilus Eurynome, Less.
——— *Eurynomus*, Jard.
Phætornis eurynomus, Gray & Mitch., Bonap.
* *Phæthornis Eurynomus*, Jard. Nat. Lib. Humming-Birds, vol. ii. p. 150.
* *Phaëtornis eurynome*, Bonap. Consp. Gen. Av. tom. i. p. 67, *Phaëtornis*, sp. 5.
* *Trochilus melanotis*, "Licht." Nordm. Erm. Reis. Atl. p. 2.
* *Phætornis melanotis*, Gray & Mitch. Gen. of Birds, vol. i. p. 104, *Phætornis*, sp. 3; Bonap. Consp. Gen. Av. tom. i. p. 67, *Phætornis*, sp. 3.
* *Ptyonornis Eurynome*, Reichenb. Aufz. der Col. p. 14; Id. Troch. Enum. p. 12.
* *Phaëthornis eurynome*, Cab. et Hein. Mus. Hein. Theil iii. p. 9.

Habitat. Brasil.

17. PHAËTHORNIS MALARIS.
Phaëthornis superciliosus . . . . Vol. I. Pl. XVII.

* *Phaëthornis malaris*, Gray & Mitch. Gen. of Birds, vol. i. p. 101, sp. 2; Bonap. Consp. Gen. Av. tom. I. p. 67, sp. 2; Cab. et Hein. Mus. Hein. Theil iii. p. 9.
* *Colibri à longue queue de Cayenne*, Buff. Pl. Enl. 600, 3.
* *Brin blanc mâle*, Vieill. Ois. Dor. tom. i. p. 37, pl. 17.
* *Trochilus superciliosus*, Id. Enc. Méth. Orn. part 2. p. 549, sp. 5; Less. Hist. Nat. des Col. p. 35, pl. 6 ; Id. Traité d'Orn. p. 288; Jard. Nat. Lib. Humming-Birds, vol. ii. p. 119, pl. 26; Cab. Schomb. Reise Guian. iii. p. 708 ; Burm. Th. Bras. ii. p. 329.
* ——— *malaris*, "Licht." Nordm. Erm. Reis. Atl. p. 2, 15.
* *Phaëthornis superciliosus*, Swains. Class. of Birds, vol. ii. p. 330; Jard. Nat. Lib. Humming-Birds, vol. ii. p. 150; Gray & Mitch. Gen. of Birds, vol. i. p. 104, *Phætornis*, sp. 1 ; Bonap. Consp. Gen. Av. tom. i. p. 67, *Phætornis*, sp. 1 ; Pelzeln, Sitz. Acad. Wien. 1856, p. 157, 1.

Habitat. Cayenne, the Guianas, and Northern Brazil?

It will be seen that the above list of synonyms differs from those given with my account of this species. I adopt them on the authority of Dr. Cabanis, who considers that I am in error in applying the term *superciliosus* to the bird I have figured under that name, and that it properly belongs to the one I have called *Pretrei*,—an opinion which is probably correct, as the German naturalists are doubtless better acquainted with the type specimens of continental writers than we can be: the synonyms of *malaris* and *superciliosus* are therefore given as stated by Dr. Cabanis.

18. PHAËTHORNIS CONSOBRINUS.

Trochilus consobrinus, "Bourc." Reichenb. Aufz. der Col. p. 17.
Phæthornis Moorei, Lawr, in Ann. Lyc. Nat. Hist. New York, vol. vi. p. 259.

Habitat. New Granada, Ecuador, and the banks of the Napo.

This is the bird so commonly sent from Bogota, and which so closely assimilates both to the *malaris* (*superciliosus* of my work) and *longirostris* (*cephalus*). It ranges over the north-western parts of Venezuela and New Granada. I have also a specimen from Archidona in Ecuador. A great number of specimens from all these countries are now before me, and among them two named *consobrinus* by M. Bourcier himself, and one from Mr. Lawrence of New York, labelled *P. Moorei*, proving that these two names have been applied to the same bird.

19. PHAËTHORNIS FRATERCULUS, *Gould* . . Vol. I. Pl. XVIII.

Habitat. Cayenne and the neighbouring countries.

Every ornithologist who has paid attention to the *Trochilidæ* must have seen a Humming-Bird from Cayenne and the adjacent countries which is very similar to, but smaller than, the *malaris* (*superciliosus* of this monograph); yet, strange to say, I find no description that will accord with it. I have therefore given it the above specific appellation. It is possible that it may be the female of the *malaris* (*superciliosus*).

20. PHAËTHORNIS LONGIROSTRIS . . . . . Vol. I. Pl. XIX.

Trochilus longirostris, De Latt.
——— *cephalus*, Bourc. et Muls., Gray & Mitch.
Phætornis cephalus, Bonap.
Ptyonornis cephalus, Reichenb.
Phaëthornis longirostris, Cab. et Hein. Mus. Hein. Theil iii. p. 9.

Habitat. Central America.

21. PHAËTHORNIS SYRMATOPHORUS, *Gould* . . Vol. I. Pl. XX.

Habitat. Ecuador.

"Irides hazel; upper mandible black; lower mandible red, tipped with black; legs and feet dark flesh-colour. Stomach contained yellow insects. All insects previously examined amongst the Humming-Birds have been black."—*Fraser in Proc. of Zool. Soc.* part 27, p. 145.

22. PHAËTHORNIS BOLIVIANA, *Gould*.

Upper mandible black; under mandible yellow, with a dark tip; above and beneath the eye a stripe of buff; chin smoky brown; throat, chest, belly, and under tail-coverts dull reddish fawn colour; crown dark brown, each feather faintly striated with buff; all the upper surface dull reddish fawn-colour, crescented with small marks of brown; base of the four outer tail-feathers on each side bronzy green, to which succeeds a bar of black, beyond which the tip is reddish buff; the two prolonged centre feathers bronze at the base, then brownish black, and white for the remainder of their length.

Total length 5¼ inches, bill 1¼, wing 2¼, tail 2¼.

Habitat. Bolivia.

This bird is somewhat allied to *P. syrmatophorus*; but it is of much smaller size and has the throat and chest differently coloured, those parts being obscure smoky grey without the conspicuous streakings of buff; the whole under-surface also, as well as the rump, is less richly coloured.

23. PHAËTHORNIS PHILIPPI . . . . Vol. I. Pl. XXI.

Trochilus Philippii, Bourc.
——— *De Filippii*, Bourc.
Phætornis Philippi, Gray & Mitch.
Phaëtornis philippi, Bonap.
Orthornis defilippi, Bonap.
Ametrornis De Filippi, Reichenb.
**Ametrornis Defilippii*, Cab. et Hein. Mus. Hein. Theil iii. p. 10, note.

Habitat. Peru or Bolivia.

24. PHAËTHORNIS HISPIDUS, *Gould* . . . Vol. I. Pl. XXII.

Trochilus (———?) *hispidus*, Gould.
Phætornis hispidus, Gray & Mitch.
Phaëtornis hispidus, Bonap.
**Pigmornis hispida*, Reichenb. Aufz. der Col. p. 14; Id. Troch. Enum. p. 12.

Habitat. Bolivia.

25. PHAËTHORNIS OSERYI . . . . Vol. I. Pl. XXIII.

Trochilus Oseryi, Bourc. et Muls.
Ametrornis Oseryi, Reichenb.
Orthornis oseryi, Bonap.
Phæthornis villosus, Lawr.

Habitat. New Granada and Ecuador.

26. PHAËTHORNIS ANTHOPHILUS . . . . Vol. I. Pl. XXIV.

Trochilus Anthophilus, Bourc.
Phætornis anthophilus, Gray & Mitch., Bonap.
Phaëtornis anthophilus, Bonap.
**Phaethornis anthophilus*, Cab. et Hein. Mus. Hein. Theil iii. p. 9.

Habitat. New Granada.

27. PHAËTHORNIS BOURCIERI . . . . . . Vol. I. Pl. XXV.

Trochilus Bourcieri, Less.
Phætornis Bourcieri, Gray & Mitch.

Phaëtornis bourcieri, Bonap.
* *Trochilus Bourcieri*, Jard. Nat. Lib. Humming-Birds, vol. ii. p. 124.
* *Phaethornis Bourcieri*, Id. p. 150.
* *Orthornis Bourcieri*, Bonap. Rev. et Mag. de Zool. 1854, p. 249.
* *Ametrornis Bourcieri*, Reichenb. Aufz. der Col. p. 14; Id. Troch. Enum. p. 12; Cab. et Hein. Mus. Hein. Theil iii. p. 10.

Habitat. Cayenne and the adjacent countries.

28. PHAËTHORNIS GUYI . . . . Vol. I. Pl. XXVI.

Trochilus Guy, Less., Jard.
Ornismya Guy, Delatt. Echo du Monde Savant, no. 45, Juin 15, 1843, col. 1069.
Phætornis Guy, Gray & Mitch.
Phaëtornis Guy, Bonap.
Trochilus apicalis, Licht., Tsch.
Phætornis apicalis, Gray & Mitch.
Phaëtornis apicalis, Bonap.
* *Phaethornis Guy*, Jard. Nat. Lib. Humming-Birds, vol. ii. p. 150; Reichenb. Aufz. der Col. p. 14; Id. Troch. Enum. p. 12.
* *Guyornis typus*, Bonap. Rev. et Mag. de Zool. 1854, p. 249.
* *Toxoteuches Guyi*, Cab. et Hein. Mus. Hein. Theil iii. p. 11.

Habitat. Trinidad.

29. PHAËTHORNIS EMILIÆ.

Trochilus Emiliæ, Bourc.
* *Phætornis Emiliæ*, Gray & Mitch. Gen. of Birds, vol. i. p. 104, *Phætornis*, sp. 7; Bonap. Consp. Gen. Av. vol. i. p. 68, *Phaëtornis*, sp. 7.

Habitat. New Granada.

On reference to my account of *P. Guyi*, it will be seen that I questioned whether the *Trochilus Emiliæ* of M. Bourcier was not identical with that species; but having since seen a great number of examples of the latter from Bogota, and of the former from Trinidad, I find that each possesses certain characters by which an experienced ornithologist would be able at once to say whence specimens of either had been received. The Andean bird, when fully adult, is rather larger in size, is much darker in general appearance, has the chin stripes less conspicuous, the apical two-thirds of the tail-feathers blacker, and their basal third and the upper tail-coverts bluish green instead of pure green. The differences in the two birds are, in fact, precisely analogous to those which occur between the *Glaucis hirsutus* and *G. affinis*.

30. PHAËTHORNIS YARUQUI . . . . . . Vol. I. Pl. XXVII.

Trochilus Yaruqui, Bourc.
* *Phaëthornis Yaruqui*, Reichenb. Aufz. der Col. p. 14; Id. Troch. Enum. p. 12.
* *Guyornis Yaruqui*, Bonap. Rev. et Mag. de Zool. 1854, p. 249.

* *Toxoteuches Yaruqui*, Cab. et Hein. Mus. Hein. Theil liL p. 11, note.
Habitat. Ecuador.

"Upper mandible black; lower deep red with a black tip; legs and feet reddish."—*Fraser, Proc. of Zool. Soc.* part 28, p. 94.

As the last three species advance in age their tails become shorter, their feathers broader, and the white fringing of the lateral ones almost obsolete.

31. PHAËTHORNIS SUPERCILIOSUS.

Phaëthornis Pretrei . . . . Vol. I. Pl. XXVIII.
* *Polytmus Cayanensis longicaudus*, Briss. Orn. tom. iii. p. 686, 13, tab. 35. fig. 5.
* *Trochilus superciliosus*, Linn. Syst. Nat. tom. i. p. 199; Lath. Ind. Orn. tom. i. p. 302; Wied, Beitr. tom. iv. p. 116; " Licht." Nordm. Erm. Reis. Atl. p. 2, 16.
*——— *Pretrei*, Delatt. et Less. Rev. Zool. 1839, p. 20.
*——— *brasiliensis*, Cab. in Schomb. Reis. Guian. tom. iii. p. 708.
* *Phætornis Pretrei*, Gray, Gen. of Birds, vol. I. p. 104, *Phætornis*, sp. 16, pl. 35; Bonap. Consp. Gen. Av. tom. i. p. 68, *Phætornis*, sp. 15.
* *Trochilus affinis*, Natt. In Mus. Vindob.
* *Phæthornis affinis*, Pelzeln, Sitz. Acad. Wien, 1857, p. 157.
——— *superciliosus*, Cab. et Hein. Mus. Hein. Theil liI. p. 9.
Habitat. Eastern Brazil.

As in the case of *P. malaris*, the above list of synonyms is given on the authority of Dr. Cabanis.

32. PHAËTHORNIS AUGUSTI . . . . . . Vol. I. Pl. XXIX.

Trochilus Augusti, Bourc.
Phætornis Augusta, Gray & Mitch.
——— *augusti*, Bonap.
Phaëtornis augustae, Bonap.
* *Phaëthornis Augusti*, Reichenb. Aufz. der Col. p. 14; Id. Troch. Enum. p. 12.
Habitat. Venezuela.

33. PHAËTHORNIS SQUALIDUS.

Phaëthornis intermedius . . . . Vol. I. Pl. XXX.
* *Trochilus squalidus*, "Natt." Temm. Pl. Col. 120. fig. 1; Less. Man. d'Orn. p. 289; Id. Hist. Nat. des Col. p. 40, pl. 8; Id. Traité d'Orn. p. 289; Jard. Nat. Lib. Humming-Birds, vol. ii. p. 125.
*——— *intermedius*, Less. Troch. p. 65, pl. 19; Jard. Nat. Lib. Humming-Birds, vol. ii. p. 123.
*——— *leucophrys*, " Licht." Nordm. Erm. Reis. Atl. p. 2, 18.

E

*Phaethornis squalidus, Jard. Nat. Lib. Humming-Birds, vol. II.
p. 151; Bonap. Comp. Gen. Av. tom. i. p. 68, sp. 11; Burm.
Th. Bras. tom. ii. p. 325.
*———, intermedius, Jard. Nat. Lib. Humming-Birds, vol. ii.
p. 150; Gray & Mitch. Gen. of Birds, vol. i. p. 104, sp. 8;
Bonap. Consp. Gen. Av. tom. i. p. 67, Phaetornis, sp. 6.
*——— leucophrys, Gray & Mitch. Gen. of Birds, vol. i. p. 104,
sp. 4.
*——— brasiliensis, Gray & Mitch. Gen. of Birds, vol. i. p. 104,
sp. 10.
*Ptyonornis intermedia, Reichenb. Aufz. der Col. p. 14; Id. Troch.
Enum. p. 12.
*Phaëthornis squalida, Cab. et Hein. Mus. Hein. Theil iii. p. 8.
Habitat. South-eastern Brazil.

We now come to that section to which Bonaparte gave the name
of *Pygmornis*. As the term implies, these birds are all extremely
diminutive; so minute, indeed, are they, that, if subjected to the
balance, their tiny bodies must be weighed by grains. That these
mites of birds perform some important office in the scale of nature
is certain, from the number both of species and individuals: they
are very widely dispersed over every part of the great country which
is inhabited by this extensive family of birds; with the exception of
one species, however (the *P. Adolphi*), they all fly to the southward
of the Isthmus of Panama. How minute must be the insects taken
by these diminutive birds, how perfect must be their vision, and how
delicately sensitive must be their tongues!

The only external difference between the sexes consists in the
longer and more graduated tails of the females; in colour they are
as nearly alike as possible.

34. PYGMORNIS LONGUEMAREUS.
Phaëthornis Longuemareus . . . . . Vol. I. Pl. XXXI.
Trochilus Longuemareus, Less.
Phaetornis Longuemareus, Gray & Mitch.
Phaëtornis Longuemareus, Bonap.
Phaëthornis Longuemari, Reich.
Trochilus Longuemareus, Jard. Nat. Lib. Humming-Birds, vol. ii.
p. 126; Cab. in Schomb. Reis. Guian. tom. iii. p. 709.
Phaëthornis Longuemarrus, Jard. Nat. Lib. vol. ii. p. 151.
Pygmornis Longuemarei, Cab. et Hein. Mus. Hein. Theil iii. p. 7,
note.
Habitat. Cayenne, Guiana, Trinidad, and the eastern part of
Venezuela.

35. PYGMORNIS AMAURA.
Phaëthornis Amaura . . . . Vol. I. Pl. XXXII.
Pygmornis Amaura, Bourc.
Phaëthornis atrimentalis, Lawr.
Pygmornis amaura, Cab. et Hein. Mus. Hein. Theil iii. p. 7, note.
Habitat. Banks of the River Napo.

36. PYGMORNIS ASPASIÆ.
Phaëthornis viridicaudata, *Gould* . . . Vol. I. Pl. XXXIII.
* *Trochilus Aspasiæ*, Bourc. et Muls. Ann. de la Soc. Linn. de Lyon, tom. iii. 1856.
* *Phaëthornis viridicaudata*, Gould, Proc. of Zool. Soc. 1857, p. 14.
* *Trochilus (Polytmus) pygmæus*, Tschudi, Consp. p. 36; Id. Faun. Per. p. 243.
* *Pygmornis viridicaudata*, Cab. et Hein. Mus. Hein. Theil iii. p. 7, note.

Habitat. Brasil and Peru.

37. PYGMORNIS ZONURA, *Gould.*
Phaëthornis zonura, *Gould* . . . . . Vol. I. Pl. XXXIV.
Habitat. Peru.

38. PYGMORNIS ADOLPHI.
Phaëthornis Adolphi, *Bourc.* . . . . . Vol. I. Pl. XXXV.
Phaëthornis Adolphi, Bourc.
Pygmornis Adolphi, Parzudaki.
* *Pygmornis Adolphi*, " Sallé, MSS." Cab. et Hein. Mus. Hein. Theil iii. p. 7, note.

Habitat. Central America.

"This," says Mr. Salvin, "is an abundant species in the forest about Yzabal, but the density of the under growth renders it extremely difficult to obtain a shot at so small and active an object. The bird is by no means shy, and takes but little notice of an observer—even searching the flowers almost within arm's reach, for the insects and honey therein contained. In movement it is extremely elegant and graceful, and, flitting from flower to flower, shows its beautifully formed tail conspicuously in every motion. Like all others of its family, it selects a small twig for its perch, giving preference to a dead one. While at rest it trims its feathers dexterously with its bill, which every now and then it cleans by rubbing it first on one side and then on the other of the twig on which it stands."—'*Ibis*,' vol. I. p. 127.

39. PYGMORNIS GRISEOGULARIS, *Gould.*
Phaëthornis griseogularis, *Gould* . . . Vol. I. Pl. XXXVI.
* *Pygmornis griseogularis*, Cab. et Hein. Mus. Hein. Theil iii. p. 8.

Habitat. New Granada; and Ecuador?

In my description of this species I have inadvertently stated that it has a crescent of black across the breast, which is not the case.

Mr. Bell of New York informs me that he has heard the "little *Pygmornis* of Panama," by which I believe the present bird is intended, "sing beautifully, the notes forming a soft, shrill, and pretty song."

40. PYOMORNIS STRIIGULARIS, *Gould.*
Phaëthornis striigularis, *Gould* . . . Vol. I. Pl. XXXVII.
* *Pygmornis striigularis,* Cab. et Hein. Mus. Hein. Theil iii. p. 7, note.
Habitat. New Granada.

41. PYGMORNIS IDALIÆ.
Phaëthornis obscura, *Gould* . . . . Vol. I. Pl. XXXVIII.
* *Trochilus Idaliæ,* Bourc. et Muls. Ann. de la Soc. Linn. de Lyon, tom. iii. 1856.
* *Phæthornis obscura,* Gould, Proc. of Zool. Soc. 1857, p. 14.
* *Pygmornis obscura,* Cab. et Hein. Mus. Hein. Theil iii. p. 7, note.
Habitat. Brazil.

42. PYGMORNIS NIGRICINCTUS.
Phaëthornis nigricinctus, *Lawr.* . Vol. I. Pl. XXXIX. fig. 1.
Phaëthornis nigricinctus, Lawr.
* *Pygmornis nigricincta,* Cab. et Hein. Mus. Hein. Theil iii. p. 7, note.
Habitat. The forests bordering the upper part of the River Amazon.

43. PYGMORNIS EPISCOPUS, *Gould.*
Phaëthornis Episcopus, *Gould* . . Vol. I. Pl. XXXIX. fig. 2.
Phaëthornis Episcopus, Gould.
* *Pygmornis episcopus,* Cab. et Hein. Mus. Hein. Theil iii. p. 7, note.
Habitat. British Guiana.

44. PYGMORNIS RUFIVENTRIS.
* *Brin blanc jeune âge,* Vieill. Ois. Dor. tom. i. p. 99, pl. 19.
* *Trochilus rufigaster,* Vieill. Nouv. Dict. d'Hist. Nat. tom. vii. p. 357; Id. Enc. Méth. Orn. part ii. p. 551.
* ———— *Davidianus,* Less. Troch. p. 50, pl. 19; Jard. Nat. Lib. Humming-Birds, vol. ii. p. 127.
* *Phæthornis Davidianus,* Jard. Nat. Lib. Humming-Birds, vol. ii. p. 131; Gray & Mitch. Gen. of Birds, vol. i. p. 108, sp. 13.
* ———— *rufigaster,* Gray & Mitch. Gen. of Birds, vol. i. p. 108, sp. 12.
* ———— *pygmæus,* Cab. in Schomb. Reis. Guian. tom. iii. p. 708.
* *Eremita Davidianus,* Reichenb. Aufz. der Col. p. 14; Id. Troch. Enum. p. 11.
* *Pygmornis Davidianus,* Bonap. Rev. et Mag. de Zool. 1854, p. 250.
* ———— *rufiventris,* Cab. et Hein. Mus. Hein. Theil iii. p. 7, note.
Habitat. Cayenne.

The above list of synonyms are given on the authority of Dr. Cabanis: it is just possible that they may refer to the female of my *P. Episcopus*; but I fear that this cannot at present be satisfactorily determined.

45. PYGMORNIS EREMITA, *Gould.*

Phaëthornis Eremita, *Gould* . . . . . . Vol. I. Pl. XL.
Trochilus Brasiliensis, Temm.
——— rufigaster, Less.
Phætornis rufigaster, Gray & Mitch.
Phaëthornis Eremita, Gould.
*Trochilus Brasiliensis, Less. Man. d'Orn. tom. ii. p. 75; Id. Traité d'Orn. p. 289.
*——— rufigaster, Jard. Nat. Lib. Humming-Birds, vol. ii. p. 89, pl. 4.
*Phæthornis rufigaster, Jard. Nat. Lib. Humming-Birds, vol. ii. p. 151; Burm. Th. Bras. tom. ii. p. 326.
*Phætornis, sp., Gray & Mitch. Gen. of Birds, vol. iii. App. p. 30a.
*Phaëtornis eremita, Bonap. Consp. Gen. Av. tom. i. p. 68, Phaëtornis, sp. 12.
*Eremita rufigaster, Reichenb. Aufz. der Col. p. 14; Id. Troch. Enum. p. 11.
*Pygmornis rufigaster, Bonap. Rev. et Mag. de Zool. 1854, p. 250.
*——— eremita, Cab. et Hein. Mus. Hein. Theil iii. p. 7.

Habitat. Northern Brazil, Bahia, the banks of the Lower Amazon.

46. PYGMORNIS PYGMÆA.

Phaëthornis pygmæus . . . . Vol. I. Pl. XLI.
Trochilus pygmæus, Spix.
Phaëtornis pygmæus, Bonap.
*Trochilus Brasiliensis, Wied, Beitr. tom. iv. p. 111.
*——— pygmæus, Burm. Th. Bras. tom. ii. p. 327.
*Eremita pygmaeus, Reichenb. Aufz. der Col. p. 14; Id. Troch. Enum. p. 10.
*Pygmornis pygmaea, Cab. et Hein. Mus. Hein. Theil iii. p. 6.

Habitat. South-eastern Brazil.

Subfamily II. TROCHILINÆ.

I commence the second volume with the *Campylopteri*, a group of Humming-Birds distinguished by their great size, by the diversity of their colouring, and by the broad dilated shafts of the first three primaries or quill-feathers of the males. The members of this group are spread over nearly the whole of the temperate regions of America, from Mexico to the equator, including Brazil, Guiana, Venezuela, and some of the West Indian Islands.

This section of the *Trochilidæ* comprises several very distinct forms:—one remarkable for a deeply forked tail, for the rich blue colouring of the body, and for the similarity in the outward appearance of the sexes; another for having the tail cuneate; while a third, comprising six or seven species, is distinguished by a very ample and rounded tail. It is for the last form alone that I have retained the generic appellation of *Campylopterus*, applying that of

Eupetomena to the deeply forked-tailed bird *macroura*, *Sphenoproctus* to the cuneate-tailed *Pampa*, and *Phæochroa* to the *Cuvieri* and the allied *Roberti*, which may be considered as aberrant, the broad shafts of the primaries (the principal characteristic of the group) being but slightly developed. These birds lead on to *Aphantochroa*.

Genus EUPETOMENA, *Gould*.

(Ε.ῦ, benè, et πτοριμη, volans.)

Generic characters.

Male.—*Bill* longer than the head, and slightly arched; *wings* moderate; *shafts* of the first two or three primaries bowed, dilated and flattened; *tail* long and dreply forked; *tarsi* partially clothed; *feet* rather small; *hind toe* shorter than the middle toe.

Female.—Similar to the male in plumage.

47. EUPETOMENA MACROURA.

Eupetomena hirundinacea . . . . Vol. II. Pl. XLII.

Trochilus macrourus, Gmel., Licht.
——— *forcipatus*, Lath.
Mellisuga Caymensis cauda bifurca, Ray, Willughb., Briss.
Ornismya hirundinacea, Less.
Polytmus macrourus, Gray & Mitch.
Prognornis macrourus, Reichenb. Aufz. der Col., p. 11; Id. Troch. Enum. p. 9, pl. 805. figs. 4873-75.
Eupetomena macroura, Bonap. Rev. et Mag. de Zool. 1854, p. 254.
Cynanthus macrourus, Jard. Nat. Lib. Humming-Birds, vol. ii. p. 149.
Eupetomena macrura, Cab. et Hein. Mus. Hein. Theil iii. p. 14.
Ornismya hirundinacea, Dev. Rev. et Mag. de Zool. 1852, p. 214.

Habitat. Brazil, Cayenne, and the neighbouring countries.

M. Deville states that "this Humming-Bird sometimes accompanies the *Chrysolampis moschitus* into the fields, but generally prefers the neighbourhood of the river-banks, where the silky tufts of the *Ingæ* and the blossoms of the numerous *Lianes* suffice for its wants. It flies very rapidly, has a shrill cry, and is so fearless that it will settle within a few feet of the object which has alarmed it. It is found throughout the whole of Brazil all the year round, but appears to be most numerous in August, September, and October."

Genus SPHENOPROCTUS, *Cab.*

Of this form there are evidently two species—one inhabiting Mexico, and the other Guatemala. It has always been considered by Trochilidists that the 15th Plate of the Supplement to 'Lesson's Histoire Naturelle des Oiseaux-mouches' represents one or other of them, but that he was in error in giving the interior of La Plata as its habitat. I have not been able to see Lesson's type; otherwise I could have ascertained to which of the two it has reference, or whether it is different from both. My figures were taken from

Guatemalan specimens, and Lesson's plate would appear to have been taken from an example procured in the same country; consequently the term *Pampa* must be retained for the Guatemalan bird, while for the larger and stouter Mexican birds we must use Lichtenstein's name *curvipennis*.

48. SPHENOPROCTUS PAMPA.

Campylopterus Pampa, *Less.* . . . . . . Vol. II. Pl. XLIII.
Ornismya Pampa, Less.
Polytmus pampa, Gray & Mitch.
Campylopterus pampa, Less., Bonap., Jard.
Pampa campyloptera, Reichenb.
* *Campylopterus pampa*, Sclat. & Salv. Ibis, vol. i. p. 127; Salv. Ibis, vol. ii. p. 260.

Habitat. Guatemala.

49. SPHENOPROCTUS CURVIPENNIS.

* *Trochilus curvipennis*, Licht. Preis-Verz. Mex. Thier. v. Deppe & Schiede (Sept. 1830), no. 32.
* *Sphenoproctus pampa*, Cab. et Hein. Mus. Hein. Theil iii. p. 11.
* *Campylopterus pampa*, Montes de Oca in Proc. Acad. Nat. Sci. Philad. 1860, p. 551.

Habitat. Mexico.

This species differs from the preceding in its much larger size, and in the paler tint of its blue crown.

"The people of Coantepec, nine miles from Jalapa," says M. Montes de Oca, "give to this species the name of *Chupa-mirto fandanguero*, or Fandango Myrtle-sucker, apparently because it has a somewhat musical voice. It is the only Humming-Bird with which I am acquainted whose notes are sufficient to recognise it by in the woods; though rather monotonous, they are very pleasing. It is occasionally found in the neighbourhood of Jalapa, but it is more abundant at Coautepec. It inhabits the forest in the winter season, and generally feeds on the flowers of the high bushes called *Asasaretos*, which are then in full bloom, and densely covered with smooth emerald-green leaves, amongst which it is very difficult to be detected. Very few are to be seen in summer time."

Genus CAMPYLOPTERUS, *Swains.*

50. CAMPYLOPTERUS LAZULUS . . . . . Vol. II. Pl. XLIV.
Trochilus lazulus, Vieill.
———— *falcatus*, Swains., Less.
Mellisuga lazulus, Gray & Mitch.
Campylopterus lazulus, Bonap.
Taeniopterus lazulus, Reichenb.
* *Ornismya falcata*, Less. Hist. Nat. des Ois. Mou. pp. xliv. 126. pl. 36.
* *Campylopterus lazulus*, Cab. et Hein. Mus. Hein. Theil iii. p. 13.

Habitat. Venezuela, the hilly parts of New Granada generally, and Ecuador, from which latter country I have received specimens through Professor Jameson, collected near Barza.

51. CAMPYLOPTERUS HEMILEUCURUS.

Campylopterus Delattrei . . . . Vol. II. Pl. XLV.

Ornismya (Campylopterus) De Lattre, Less.
Mellisuga De Lattrei, Gray & Mitch.
Campylopterus delattre, Bonap.
———————— *delattrii,* Bonap.
———————— *Delattrei,* Reichenb.
* *Trochilus hemileucurus,* Licht. Preis-Verz. Mex. Thier. v. Deppe & Schiede (Sept. 1830), no. 33.
* *Campylopterus hemileucurus,* Cab. et Hein. Mus. Hein. Theil iii. p. 13.
* ———————— *De Lattrei,* Montes de Oca in Proc. Acad. Nat. Sci. Philad. 1860, p. 47.

Habitat. Mexico and Guatemala.

"The large and showy tail of this Humming-Bird," says Mr. Salvin, "makes it one of the most conspicuous when on the wing. It is common at Coban, feeding among the *Salvia*; it is said also to be found in the Volcan de Fuego, but I have not met with it. The females of this species are most abundant, their ratio to the males being as five to two. It is not nearly so shy as its congener, *C. rufus.*"—*Ibis,* vol. II. p. 260.

"This beautiful Humming Bird," says M. Montes de Oca, "is generally known in Mexico by the name of *Chupa-mirto real azul,* or Royal Blue Myrtle-sucker. It arrives in the vicinity of Jalapa, Coantepec, and Orizaba in considerable numbers during the months of October and November, and is mostly found feeding from a plant called *Masnpan,* between the hours of nine and one o'clock. During this time it is seldom seen to alight, and then only for a very short time in any one place, but is constantly on the wing, flitting from flower to flower, describing the segment of a circle in its flight, and sometimes almost touching the ground. For the remainder of the day very few are to be seen, and I think it probable that they visit the woods for certain kinds of mosquitoes, with which I have often found their stomachs well filled.

"The pugnacity of this species is very remarkable. It is very seldom that two males meet without an aërial battle. The contest commences with a sharp choleric shriek, after which, with dilated throats, the feathers of the whole of their bodies erected on end, and their tails outspread, they begin to fight with their bills and wings, and the least powerful soon falls to the ground or flies away. I have never known one of these battles last longer than about ten seconds; and in the specimens I have had under my notice in cages, their fighting has mostly ended in the splitting of the tongue of one of the two, which then surely dies from being unable to feed."

53

52. CAMPYLOPTERUS ENSIPENNIS . . . . Vol. II. Pl. XLVI.
Trochilus ensipennis, Swains.
Campylopterus ensipennis, Less., Jard., Bonap., Reichenb.
Polytmus ensipennis, Gray & Mitch.
* *Trochilus latipennis*, Jard. Nat. Lib. Humming-Birds, vol. L
  p. 116, pl. 34.
* *Campylopterus latipennis*, Jard. ib. p. 153.
* ————— *ensipennis*, Cab. et Hein. Mus. Hein. Theil iii. p. 12.
Habitat. The Island of Tobago.

On reference to my account of *Campylopterus Villavicencio*, it will be seen that I was inclined to believe the *C. splendens* of M. Lawrence to be identical with that bird; but on reconsidering the matter, and observing how numerous and how closely allied are the species of the genus *Campylopterus*, I now think it probable that it is really distinct. The throat in *C. splendens* is beautiful blue, and the abdomen washed with green; while in *C. Villavicencio*, the whole of the under surface is pure grey. Both these birds have fine metallic-green crowns, which circumstance induced me to believe that they were opposite sexes of one and the same species, and it is possible that they may yet prove to be so; but for the present I shall regard them as distinct.

53. CAMPYLOPTERUS SPLENDENS, *Lawr.* . Vol. II. Pl. XLVII.
(Upper fig.)
Campylopterus splendens, Lawr. in Ann. Lyc. Nat. Hist. New York, vol. vi. p. 262.
Habitat. The forests between the upper waters of the Napo and Quito.

54. CAMPYLOPTERUS VILLAVICENCIO . . Vol. II. Pl. XLVII.
(Lower fig.)
Trochilus Villaviscensio, Bourc.
Heliomaster Villavicensio, Reichenb.
Heliomastes villavisencio, Bonap.
Habitat. Forests bordering the Rio Napo in Ecuador.

55. CAMPYLOPTERUS LATIPENNIS. . . . . Vol. II. Pl. XLVIII.
Trochilus campylopterus, Gmel., Valenc., Draplez.
————— *cinereus*, Gmel., Lath.
————— *largipennis*, Bodd.
————— *latipennis*, Lath., Vieill., Swains., Jard.
Polytmus largipennis, Gray & Mitch.
Ornismya latipennis, Less.
Campylopterus latipennis, Swains., Jard., Bonap., Less., Reichenb.
* *Campylopterus latipennis*, Cab. in Schomb. Reis. Guian. tom. iii. p. 709.
* ————— *largipennis*, Cab. et Hein. Mus. Hein. Theil iii. p. 12.
Habitat. Cayenne and British Guiana.

In my account of *C. latipennis* I have stated my belief that

Genus DOLERISCA, *Cab.*

The typical species of this form is the *Trochilus fallax* of M. Bourcier, a bird distinguished by its tawny-coloured breast, and by the white tippings of its outer tail-feathers. I wish it to be understood that I do not include in this genus the *albicollis* or the *chionogaster*, which have been inadvertently figured as pertaining to it. At the same time were I to state that the genus is confined to a single species, I believe that I should be leading ornithologists into error; for I have a specimen which, I think, will prove to belong to a second. The example in question, although bearing all the general characteristics of the *T. fallax*, differs in some minor details, and I shall therefore provisionally propose for it the specific name of *cervina*.

64. DOLERISCA FALLAX.

Leucippus fallax . . . . Vol. II. Pl. LVI.

Trochilus fallax, Bourc.
——— (*Lampornis?*) *fulviventris*, Gould.
Polytmus fallax, Gray & Mitch.
Leucippus fallax, Bonap., Reichenb.
Doleromyia fallax, Bonap.
Dolerisca fallax, Cab. et Hein. Mus. Hein. Theil iii. p. 6.

Habitat. Venezuela.

65. DOLERISCA CERVINA, *Gould.*

Habitat. Unknown.

This new species is larger than the *fallax* in all its admeasurements, and has a lesser amount of white on the tips of the outer tail-feathers. In *fallax* these greyish-white tippings occupy both webs of the apical portion of each of the three outer feathers, while in the *cervina* the inner webs only are thus marked; these marks are about three-eighths of an inch long on the outer feather, a quarter of an inch on the next, and but a little more than an eighth on the third; the upper mandible in *cervina* is reddish brown, while in *fallax* the upper one is black. The habitat of the latter is well known to be Venezuela, but that of the former has yet to be ascertained.

Genus UROCHROA, *Gould.*

(Οὐρά, cauda, et χρόα, color.)

Generic characters.
Male.—*Bill* lengthened and straight, or slightly arched; *wings* moderately long and pointed; *tail* square; *tarsi* partly clothed; *hind toe* as long as the middle toe; *nails* short.

Female.—Unknown.

Of this remarkable form only one species is at present known.

66. UROCHROA BOUGUERI . . . . Vol. II. Pl. LVII.

Trochilus Bougueri, Bourc.

Cæligena bougueri, Bonap.
Cæligena Bouguieri, Reichenb.
Urochroa bougieri, Sclat. Proc. Zool. Soc. part 28. p. 95.
Habitat. Nanegal, in Ecuador.

Genus STERNOCLYTA, *Gould*.
(Στέρνον, pectus, et ἀυτὸs, insignis.)

Generic characters.

Male.—*Bill* unusually large, rather arched, and much longer than the head; *wings* ample; *tail* moderate and rounded; *tarsi* partly clothed; *feet* moderate; *throat* and *breast* luminous.
Female.—Unadorned.
Three outer tail-feathers tipped with white in both sexes.

67. STERNOCLYTA CYANEIPECTUS, *Gould* . Vol. II. Pl. LVIII.

Trochilus (*Lampornis*) *cyanopectus*, Gould.
Sternoclyta cyanopectus, Gould.
Campylopterus cyanipectus, Bonap.
Lampornis cyanopectus, Bonap.
Sarpiopterus cyanopectus, Reichenb.
* *Polytmus cyanopectus*, Gray & Mitch. Gen. of Birds,vol. i. p. 108.
Polytmus, sp. 22.
* *Sternoclyta cyanipectus*, Cab. et Hein. Mus. Hein. Theil iii. p. 13, note.

Habitat. The province of La Guayra in Venezuela.

We now proceed to the genera *Delatiria, Cæligena, Lamprolæma, Eugenes,* and their allies, all of which are peculiar to Central America; at least, so far as is yet known, none of them have been found to the southward of the Isthmus; even Veragua, so far as we are aware, is not tenanted by any one of them.

It may be considered by some ornithologists that here the subdivision of genera has been carried too far; but having once broken ground, and separated the old genus *Trochilus*,it would be inconsistent to place together in one genus all the members of this Central American group of Humming-Birds; for while a certain degree of unity pervades them, no generic character could be found which would be applicable to the whole. This instance will serve most efficiently to illustrate the great diversity of closely allied forms which occur in the great family of Humming-Birds. We frequently find groups, like the present, so diversified that nearly every species demands a generic title, while in such genera as *Thalurania, Petasophora,* and *Aglæactis,* the species, though as distinct as they well can be, possess characters common to all.

I commence with the

Genus EUGENES, *Gould*.
(Εὐγενής, nobilis.)

Generic characters.

Male.—*Bill* straight, longer than the head; *wings* long and pointed; *tail* moderate and very slightly forked; *tarsi* clothed; *feet* rather small; *hind toe* about equal in length to the middle one; *crown* and *throat* luminous.

Female.—Unadorned.

Of this form only one species is known; it is a native of Guatemala and Southern Mexico, and is distinguished from its allies by the gorgeous colouring of its crown and breast. It is in the possession of a luminous crown, and other characters, that this bird differs from that immediately following.

68. EUGENES FULGENS. . . . . Vol. II. Pl. LIX.

Trochilus fulgens, Swains.
Ornismya Rivolii, Less.
Trochilus Rivolii, Jard.
Mellisuga fulgens, Gray & Mitch.
Delattria fulgens, Bonap.
Cœligena fulgens, Bonap.
Cœligena fulgens, Reichenb.
*Cœligena fulgens, Reichenb. Troch. Enum. p. 3, pl. 686. figs. 4513–14.
*Eugenes fulgens, Cab. et Hein. Mus. Hein. Theil iii. p. 20.
**Trochilus melanogaster*, Licht. in Mus. Berlin.
*——— *Rivoli*, Swains. Birds of Brazil, pl. 76.

Habitat. Mexico and Guatemala.

"This species," says Mr. Salvin, "is rare at Coban. The western boundary of the Llano of Dueñas is the spot where I have found it in the greatest numbers; indeed, with two exceptions, I have never met with it elsewhere. It is a most pugnacious bird. Many a time have I thought to secure a fine male, which I had perhaps been following from tree to tree, and had at last seen quietly perched on a leafless twig, when my deadly intention has been anticipated by one less so in fact, but to all appearance equally so in will. Another Humming-Bird rushes in, knocks the one I covet off his perch, and the two go fighting and screaming away at a pace hardly to be followed by the eye. Another time this flying fight is sustained in mid air, the belligerents mounting higher and higher, till the one worsted in battle darts away, seeking shelter, followed by the victor, who never relinquishes the pursuit till the vanquished, by doubling and hiding, succeeds in making his escape. These fierce raids are not waged alone between members of the same species. *Eugenes fulgens* attacks with equal ferocity *Amazilia dumerilii*, and, animated by no high-souled generosity, scruples not to tilt with the little *Trochilus colubris*. I know of hardly any species that shows itself more brilliantly than this when on the wing; yet it is not to the midday sun that it exhibits its splendour. When the sultry wind brings clouds and driving mist between the volcanos of Agua and Fuego, and all is as in a November fog in England, except that the yellow element is wanting, then it is that *Eugenes fulgens* appears in numbers; *Amazilia Devillei*, instead of a few scattered birds, is to be seen in every tree, and *Trochilus colubris* in great abundance: such animation awakes in Humming-Bird life as would hardly be credited by one who had passed the same spot an hour or two before; and the flying to and fro, the humming of wings, the momentary and prolonged contests, and the incessant battle-cries seem almost enough for a time to turn the head of a lover of these things. I have fifteen males from Dueñas to one female."—*Ibis*, vol. ii. p. 261.

Following the *Eugenes fulgens* is the softly coloured *Delattria Clemenciæ* of my work, the proper name of which is *Cæligena Clemenciæ*, it being the type of the

Genus CŒLIGENA, Less.

69. CŒLIGENA CLEMENCIÆ, Less.

Delattria Clemenciæ . . . . Vol. II. Pl. LX.
Ornismya Clemenciæ, Less.
Lampornis Clemenciæ, Less.
Cæligena Clemenciæ, Less.
Mellisuga Clemenciæ, Gray & Mitch.
Delattria clemencia, Bonap.
Lampornis clemencia, Bonap.
Cæligena Clemencia, Reichenb.
* *Campylopterus Clemenciæ*, Jard. Nat. Lib. Humming-Birds, vol. ii. p. 154.
* *Trochilus lucidus*, Licht. in Mus. of Berlin.
* *Cæligena Clemenciæ*, Reichenb. Troch. Enum. p. 3, pl. 687. fig. 4516; Cab. et Hein. Mus. Hein. Theil iii. p. 15.
Habitat. Mexico, where it is far from common.

The
Genus LAMPROLÆMA, Reichenb.

was instituted for the truly beautiful bird known as De Rham's Garnet.

70. LAMPROLÆMA RHAMI . . . . . . Vol. II. Pl. LXI.

Ornismya Rhami, Less.
Ornismia Rhami, Delatt. et Less.
Mellisuga Rhami, Gray & Mitch.
Lampornis rhami, Bonap.
Delattria rhami, Bonap.
Lamprolaima Rhami, Reichenb.
Heliodoxa Rhami, Reichenb.
* *Trochilus fulgidus*, Licht. in Mus. of Berlin.
* *Lamprolæma Rhami*, Cab. et Hein. Mus. Hein. Theil iii. p. 30.
Habitat. Guatemala.

We now come to the

Genus DELATTRIA, Bonap.

as restricted to the *D. Henrici* and *D. viridipallens*, both of which species are natives of Guatemala.

71. DELATTRIA HENRICI . . . . . . Vol. II. Pl. LXII.

Ornysmia Henrica, Less. et Delatt.
Topaza Henrica, Gray and Mitch.
Delattria henrica, Bonap.
―――― *henrici*, Bonap.
Lamprolaima Henrici, Reichenb. Aufz. der Col. p. 9.
Heliodoxa Henrici, Reichenb. Troch. Enum. p. 6, pl. 742. figs. 4701–3.

Habitat. Guatemala.

72. DELATTRIA VIRIDIPALLENS . . . . Vol. II. Pl. LXIII.

Trochilus viridi-pallens, Bourc. et Muls.
Polytmus viridi-pallens, Gray & Mitch.
Delattria viridi-pallens, Bonap.
Thaumantias viridipallens, Bonap.
Agyrtria viridipallens, Reichenb.

Habitat. Guatemala.

"Occurs, in company with *Petasophora thalassina*, on the Volcan de Fuego. Seems to keep entirely to the forests of the volcano. I have never met with it in the plains below. This is one of the commonest species at Coban. It may readily be recognized by the peculiar harshness of its note."—*Salvin in 'Ibis,'* vol. ii. pp. 40, 263.

Near to these are the members of the

Genus HELIOPÆDICA, *Gould*.

(Ἥλιος, sol, et παιδικὸs, juvenilis.)

Generic characters.

Male.—Bill straight, and rather longer than the head; *head* round, or with the feathers not advancing on the bill; *tail* slightly rounded, the feathers broad; *tarsi* clothed; *hind toe* shorter than the middle one; *head* and *breast* luminous.
Female.—Unadorned.

This genus comprises two species, both of which are natives of Central America, Mexico, and Southern California; they are somewhat diminutive in size, and possess the white mark behind the eye which occurs in most of the members of the genera of this section of the *Trochilidæ*.

73. HELIOPÆDICA MELANOTIS . . . . Vol. II. Pl. LXIV.

Trochilus melanotus, Swains.
Ornismya Arsenni, Less.
Trochilus leucotis, Vieill.?
Thaumatias leucotis, Bonap.?
Basilinna leucotis, Reichenb.?
Trochilus leucocrotaphus, Shaw (Cabanis).
* ―――― *cuculliger*, Licht., Preis-Verz. Mex. Thier. v. Deppe & Schiede (Sept. 1830), no. 29, 31.

Trochilus leucotis, Jard. Nat. Lib. Humming-Birds, vol. ii. p. 144.
Hylocharis leucotis, Gray & Mitch. Gen. of Birds, vol. i. p. 114, *Hylocharis*, sp. 28.
Heliopædica melanotis, Sclat. & Salv. Ibis, vol. i. p. 130.
Basilinna leucotis, Cab. et Hein. Mus. Hein. Theil iii. p. 45.
Trochilus lucidus, Shaw? Gen. Zool. vol. viii. p. 327.
Mellisuga lucida, Steph. Cont. of Shaw, Gen. Zool. vol. xiv. p. 247.
Sapphironia lucida, Sallé, Liste des Oiseaux; Sclat. Proc. Zool. Soc. part xxvi. p. 297, and part xxvii. p. 386.

Habitat. Guatemala and Mexico.

I observe that specimens from Guatemala are much smaller than those from Mexico; but as the colouring and disposition of the markings are precisely similar, I regard them as races only.

M. Sallé, in his 'List of the Birds of Mexico,' has assigned to one of them the name of *lucidus* of Shaw, believing it to be an earlier name for this bird than *melanotis* or *Arsenni*. This list has been followed by Dr. Sclater in his papers on the birds received by M. Sallé from and collected by M. Boucard in Oaxaca; but as Shaw's description of *lucidus*, as well as the country in which it is said to be found (Paraguay), does not accord with that of *melanotis*, that name must sink into a synonym.

"In some of the open savannahs scattered among the oak-forests of the Volcan de Fuego near Calderas, this species is not uncommon; I have also frequently met with it in some of the 'barrancos' of the same volcano. The white mark running from the eye and the deep coral-red of the bill show conspicuously in the living bird. It is a very shy species. A single specimen was shot near Coban, and another was brought to me from the mountains of S. Cruz, near San Gerónimo."—*Salvin in Ibis*, vol. ii. p. 271.

74. HELIOPÆDICA XANTUSI . . . . Vol. II. Pl. LXV.

Amazilia Xantusii, Lawr.
Heliopædica castaneocauda, Lawr.
Habitat. Southern California.

If I have extolled the members of the genus *Cometes* as being among the most gorgeous birds in existence with regard to the colouring of their tails, in like manner I may pronounce the *Topazæ*, which now claim our notice, to be as remarkable for their lustrous throat-marks.

One of these beautiful birds, the *Topaza Pella*, is an inhabitant of Cayenne and the adjacent countries; while another, the *T. Pyra*, flies in the forests of the Upper Rio Negro.

Genus TOPAZA, *G. R. Gray.*

75. TOPAZA PELLA . . . . Vol. II. Pl. LXVI.

Polytmus Surinamensis longicaudus ruber, Briss.
Trochilus pella, Linn. et Auct.

Falcinellus guttore viridi, Klein.
Certhia Surinamensis, Spalowsky.
Colibri pella, Less.
Topaza pella, Gray & Mitch., Bonap., Reich., Cabanis.
* *Trochilus paradiseus*, Linn. Syst. Nat. tom. i. p. 189.
* *Lampornis pella*, Jard. Nat. Lib. Humming-Birds, vol. ii. p. 155.
Habitat. Cayenne and the adjacent countries.

I find that specimens from Demerara have more-richly coloured throat-marks than those procured in Cayenne; there is also another variety distinguished by the great breadth of their lengthened tail-feathers; but these differences are not of specific importance.

76. TOPAZA PYRA . . . . Vol. II. Pl. LXVII.

Trochilus (Topaza) pyra, Gould.
Topaza pyra, Gray, Bonap., Reichenb., Cabanis.
Habitat. The Upper Rio Negro.

It is only at a comparatively recent date that we became acquainted with the birds for which I proposed the term *Oreotrochilus*. D'Orbigny introduced to us the *O. Estellæ* and *O. Adelæ*; while in 1846 the fine *O. Chimborazo* was brought to light through the researches of M. Bourcier; in 1849 the same gentleman made us aware of the existence of the little less beautiful *O. Pichincha*, and I, on my own part, had the pleasure of making known the *O. melanogaster* and *O. leucopleurus*. All these birds inhabit loftier elevations than any other genus of Humming-Birds; for they love to dwell in regions just beneath the line where the melting snows and the warmth of the sun call forth an alpine flora and a peculiar character of insect life; and I question if any other insessorial birds seek their food at so great an elevation as the *O. Chimborazo* and *O. Pichincha*. As far as our present knowledge extends, no species has been found to the northward of Ecuador, while to the south they range along the highlands of Peru and Bolivia.

Genus OREOTROCHILUS, Gould.

("Opos, mons, et τρόχιλος, trochilus; Mountain Humming-Bird.)

Generic characters.

Male.—Bill longer than the head, almost cylindrical, and slightly incurved; *wings* rather long and powerful; *tail* large, the feathers narrow and rigid; *tarsi* clothed; *feet* strong; *hind toe* and *nail* about the same length as the middle toe and nail; *throat* luminous.

Female.—Unadorned.

77. OREOTROCHILUS CHIMBORAZO . . . Vol. II. Pl. LXVIII.

Trochilus Chimborazo, Bourc.
Oreotrochilus Chimborazo, Gould, Gray & Mitch., Bonap., Reich.
* *Orotrochilus Chimborazo*, Cab. et Hein. Mus. Hein. Theil iii. p. 15, note.

Habitat. Immediately below the snow-line round the cone of the volcanic mountain Chimborazo.

Mr. Fraser, who killed many examples at Panza, at an altitude of 14,000 feet, says, "Irides hazel: bill, legs, and feet black. To be seen occasionally on the *Arbor Maria*, but feeds generally on a red thistle. It is common, and by no means shy, and has rather a pretty song, oft repeated, and to be heard at a considerable distance. In bad weather, when the wind is high, this bird is said to creep under and into the clumps of *Paja* (a species of *Stipa*)."

78. OREOTROCHILUS PICHINCHA . . . . Vol. II. Pl. LXIX.
Trochilus Pichincha, Bourc. et Muls.
Oreotrochilus Jamesoni, Jard.
—————— *Pichincha*, Bonap., Reichenb.
?*Orotrochilus Pichinchae*, Cab. et Hein. Mns. Hein. Theil iii. p. 15.
Habitat. The snow-line of the volcanic mountains of Pichincha and Cotopaxi in Ecuador.

"Guagua and Rueo Pichincha (14,000 feet alt.), many examples. The Pichincha Humming-Bird, like the Chimborazo, is found only close under the line of perpetual snow; but this species, according to the present state of our knowledge, is more widely distributed than the latter, being found not only on Pichincha, but also on Antisana and Cotopaxi. Upon my first visit to Guagua Pichincha these birds were feeding entirely on the ground, hunting the little moss-covered clumps as fast as the snow melted. They are not uncommon in this locality, but always met with singly. They are very restless, but not shy, seldom remaining on one clump more than a second, then away to another, perhaps a yard distant. Sometimes they would take a rapid flight of 40 or 50 yards. On my second visit, the Chuquiragua (*Chuquiraga insignis*, Humb.) being in flower, they were feeding from it like the *Quindi* of Chimborazo, but still occasionally hunted the mossy clumps. They flit with a *burr* of the wings, and occasionally settle, with the feathers all ruffled, on the top of the Chuquiragua or other small plant. In this respect, so far as my observations and those of Professor Jameson go, they differ from *O. Chimborazo*.

"June 5. No snow on the ground, and all birds were apparently scarce and shyer; these birds in particular were chasing each other, in twos and threes, like flashes of lightning."—*Fraser in Proc. of Zool. Soc.* part xxviii. p. 79.

79. OREOTROCHILUS ESTELLA . . . . . Vol. II. Pl. LXX.
Trochilus Estella, D'Orb. et La Fres.
Orthorhynchus Estella, D'Orb.
Trochilus Ceciliæ, Less.
Oreotrochilus Estella, Gould, Gray & Mitch., Bonap., Cabanis.
?*Orotrochilus Estellae*, Cab. et Hein. Mus. Hein. Theil iii. p. 16.
Habitat. The high lands near La Paz in Bolivia.

80. OREOTROCHILUS LEUCOPLEURUS, *Gould* . Vol. II. Pl. LXXI.
Oreotrochilus leucopleurus, Gould, Gray & Mitch., Bonap., Reichenb.

* *Orotrochilus leucopleurus*, Cab. et Hein. Mus. Hein. Theil iii. p. 16.
* " *Oreotrochilus leucopterus*, Reichenb.", Cab. et Hein. ib.
* *Trochilus Milleri*, Lodd. MS.; Fras. in Proc. of Zool. Soc. part xi. p. 114.

Habitat. The Chilian Andes.

"This beautiful and rare species of Humming-Bird," says Mr. Bridges, "is only found in the elevated valleys of the Andes, residing amongst storms of hail, rain, and thunder, and in places where the naturalist would least expect to find a species of *Trochilus*. It subsists more upon small flies than upon the nectar of flowers. On examination of the crops I found them filled with flies, which they take before sun-down along the margin of the mountain rivulets. Specimens were taken at Los Ojos de Aqua, province of Aconcagua, at an elevation of from 6000 to 8000 feet, and I saw them at least 1000 feet above that place. Iris brown."—*Proc. Zool. Soc.* part xi. p. 114.

Dr. Philippi met with this bird at Hueso Parado in Northern Chili, at an elevation of not more than 1000 feet above the sea-level.

81. OREOTROCHILUS MELANOGASTER, *Gould*. Vol. II. Pl. LXXII.

Oreotrochilus melanogaster, Gould, Gray & Mitch., Bonap., Reichenb.
Orotrochilus melanogaster, Cab. et Hein. Mus. Hein. Theil iii. p. 15, note.

Habitat. The high lands of Peru; precise locality unknown.

82. OREOTROCHILUS ADELÆ . . . . . Vol. II. Pl. LXXIII.

Trochilus Adela, D'Orb. et Lafresn.
Orthorhynchus Adela, D'Orb.
Oreotrochilus Adelæ, Gould, Gray & Mitch., Bonap., Reichenb.
* *Orotrochilus Adelae*, Cab. et Hein. Mus. Hein. Theil iii. p. 15, note.

Habitat. Bolivia; the high lands around Chuquesaca being one of its localities.

I now proceed to the

Genus LAMPORNIS, *Swains.*

This genus comprises many species, some of which inhabit the West Indian Islands, and others the mainland. The best-known among them, the *Lampornis Mango*, has a wider range than any of the others, as will be seen on reference to my account of that species. They are all distinguished by the harmonious colours of their ample tails, which are even more beautiful in the females than in the males.

83. LAMPORNIS MANGO . . . . Vol. II. Pl. LXXIV.

Trochilus Mango, Linn. et auct.
——— *violicauda*, Bodd.
——— *albus*, Gmel.
——— *punctulatus*, Gmel.

Trochilus nitidus, Lath.
Polytmus punctulatus, Briss.
Trochilus atricapillus, Vieill.
——— *fasciatus*, Shaw.
——— *quadricolor*, Vieill.
——— *nigricollis*, Vieill.
Lampornis Mango, Swains., Bonap.
Polytmus Mango, Gray & Mitch.
Anthracothorax Mango, Reichenb.
* *Trochilus punctatus*, Vieill. Ency. Méth. Orn. part ii. p. 550 (young).
* ——— *lazulus*, Less. Traité d'Orn. p. 290.
* *Lampornis Mango*, Cab. et Hein. Mus. Hein. Theil iii. p. 19.

Habitat. The eastern part of Brazil, Trinidad, Venezuela, and the high lands of New Granada.

64. LAMPORNIS IRIDESCENS, *Gould*.

This is the bird from Guayaquil which I have spoken of in my account of *L. Mango* as differing from the Mangos of the other parts of America. The chief differences are a rather shorter tail and a glittering wash of blue and green on the throat, instead of that part being velvety black; there is also a greater amount of green on the flanks. Three specimens of this bird were killed and sent to me by Professor Jameson during one of his visits to the coast.

Habitat. Guayaquil.

65. LAMPORNIS PREVOSTI . . . . . . Vol. II. Pl. LXXV.

Trochilus Prevostii, Less., Bourc.
Polytmus Prevostii, Gray & Mitch.
Lampornis prevosti, Bonap.
Anthracothorax Prevostii, Reichenb.

Habitat. Guatemala and Honduras.

66. LAMPORNIS VERAGUENSIS, *Gould* . . Vol. II. Pl. LXXVI.

Lampornis Veraguensis, Gould, Bonap.
Sericotes Veraguensis, Reichenb.
Anthracothorax Veraguensis, Reichenb. Troch. Enum. p. 9, pl. 799. fig. 4848.
Lampornis Veraguensis, Cab. et Hein. Mus. Hein. Thiel iii. p. 18.

Habitat. Veragua and Costa Rica.

Mr. Bridges "found this species in the outskirts of the town of David, feeding among the flowers of a large arborescent species of *Erythrina*."

67. LAMPORNIS GRAMINEUS . . . . . Vol. II. Pl. LXXVII.

Trochilus gramineus, Linn. et auct.
——— *pastorulis*, Lath., Vieill., Steph.
——— *maculatus*, Gmel., Vieill.
——— *gularis*, Gmel., Lath., Vieill.

Trochilus marmoratus, Vieill.
Polytmus dominicus, Gray & Mitch., Bonap.
Lampornis dominicus, Bonap.
Hypophania dominica, Reichenb.
Anthracothorax dominicus, Reichenb. Troch. Enum. p. 9, pl. 792, figs. 4845–46.
Lampornis graminea, Cab. et Hein. Mus. Hein. Theil iii. p. 18.

Habitat. Trinidad, Cayenne, and Guiana.

88. LAMPORNIS VIRIDIS . . . . . . Vol. II. Pl. LXXVIII.

Trochilus viridis, Aud. et Vieill., Bonu., Dumont, Drapiez, Temm.
Le Colibri cyaneure, Trochilus viridis, Less.
Chalybura viridis, Reichenb. Aufz. der Col. p. 10.
Agyrtria viridis, Reichenb. Troch. Enum. p. 7, pl. 765, figs. 4771–72.

Habitat. Porto Rico.

89. LAMPORNIS AURULENTUS . . . . . Vol. II. Pl. LXXIX.

Trochilus aurulentus, Vieill. et auct.
Polytmus aurulentus, Vieill.
——————— *margaritaceus*, Gray & Mitch.
Lampornis margaritaceus, Bonap.
Eulampis aurulentus, Bonap.
Margarochrysis aurulenta, Reichenb.
Trochilus dominicus, Linn., Gmel., Lath., female?
Polytmus dominicus, Briss., female?

Habitat. St. Domingo.

90. LAMPORNIS VIRGINALIS, *Gould* . . . Vol. II. Pl. LXXX.

Crown and all the upper surface bronzy green; wings light purplish brown, shining greenish wax-yellow; chest and centre of the abdomen black, passing into green on the flanks; upper tail-coverts brilliant bronzy green; two centre tail-feathers rich bronze, the remainder fine purple; margined and tipped with bluish black; bill black; feet dark brown.

Total length 4⅛ inches; bill ⅝; wing 2¾; tail 1⅜; tarsi ¼.

Habitat. The Island of St. Thomas.

If I have led my friend, Alfred Newton, Esq., into an error, by causing him to state in the 'Ibis,' vol. i. p. 375, that the *Lampornis aurulentus* is found in the Island of St. Thomas: it was quite unintentional on my part. Since we made an examination and comparison of specimens of *aurulentus* from St. Domingo, with those, which we believed to be identical, from St. Thomas, I have received numerous other examples from the latter island, a careful consideration of which induces me to regard them as distinct; and as such, I have described them under the name of *Lampornis virginalis*. The difference between this new species and *aurulentus* is very marked: it is of much smaller size, and has a shorter, more square, and differently coloured tail, the two centre feathers being rich bronze instead of purplish black; the throat-mark is richer; the upper tail-coverts are very much finer and more brilliant; and the bill is shorter.

91. LAMPORNIS PORPHYRURUS . . . . . Vol. II. Pl. LXXXI.

Trochilus porphyrurus, Shaw, Steph.
——— *bromicolor*, Less.
——— *Floresii*, Bourc.
Polytmus porphyrurus, Gray & Mitch.
Lampornis Mango, Gosse.
——— *porphyrurus*, Bonap.
——— *floresi*, Bonap.
Floresia porphyrura, Reichenb.
**Anthrocothoras porphyrurus*, Reichenb. Troch. Enum. p. 8, pl. 794. figs. 4849–50.
**Lampornis porphyrura*, Cab. et Hein. Mus. Hein. Theil iii. p. 19.

Habitat. Jamaica.

This species differs from all its allies in the female or the young male assimilating to the male in the colour of the tail, which is quite contrary to what occurs in the females of the other species; unlike them also, the female of this species has a different and more beautiful gorget than the male. This is one of the anomalies which cannot be explained, inasmuch as in structure, in size, and other characters it is a true *Lampornis*.

The genus *Eulampis* now claims our attention. It is composed of four species, the distinguishing features of which are their luminous upper tail-coverts. These broad and glittering feathers, resembling plates of shining metal, have doubtless been designed for no special purpose connected with the habits of the bird, but for mere ornament; but such characters, trifling though they be, are of no little use in enabling us to group together nearly allied species. It will be recollected that in some genera—that of *Hypuroptila* for instance—the under and not the upper tail-coverts are extraordinarily developed; and many other instances might be cited of a similar development of other parts of the plumage, for which no other use but that of mere ornament can be conceived. The members of this genus differ from most others in the perfect similarity in the colouring of the sexes. So far as I am aware, they are all confined to the West Indian Islands.

Genus EULAMPIS, *Boie.*

92. EULAMPIS JUGULARIS . . . . . . Vol. II. Pl. LXXXII.

Trochilus jugularis, Linn., Gmel., Lath., Temm.
Eulampis jugularis, Bonap., Reichenb.
Polytmus jugularis, Gray & Mitch.
Trochilus auratus, Gmel., Less.
——— *granatinus*, Lath.
——— *Bancrofti*, Lath.
——— *cyanomelas*, Gmel.

Trochilus violaceus, Gmel.
———— *auritus*, Vieill.
Polytmus Cayennensis violaceus, Briss.
Topaza violacea, Gray & Mitch.
Certhia prasinoptera, Lath., Sparrm.
Cyanathus ? jugularis, Jard.
Trochilus cyaneus, Lath.
———— *venustissimus*, Gmel.
**Eulampis jugularis*, Cab. et Hein. Mus. Hein. Theil iii. p. 17.
Habitat. The Islands of Nevis and Martinique.

93. EULAMPIS HOLOSERICEUS . . . . Vol. II. Pl. LXXXIII.

Trochilus holosericeus, Linn. et auct.
Polytmus mexicanus, Briss.
Trochilus aurigaster, Shaw.
Polytmus holosericeus, Gray & Mitch.
Eulampis holosericeus, Bonap.
Sericotes holosericeus, Reichenb.
**Anthracothorax holosericeus*, Reichenb. Troch. Enum. p. 9, pl. 793. fig. 4847.
**Trochilus atrigaster*, "Shaw," Cabanis.
**Eulampis holosericea*, Cab. et Hein. Mus. Hein. Theil iii. p. 17.
Habitat. Islands of Nevis? and Martinique?

94. EULAMPIS CHLOROLÆMUS, *Gould* . . Vol. II. Pl. LXXXIV.

Sericotes chlorolaimus, Reichenb.
Eulampus chlorolæmus, Bonap.
**Anthrocothorax chlorolaimus*, Reichenb. Troch. Enum. p. 9.
**Eulampis chlorolaema*, Cab. et Hein. Mus. Hein. Theil iii. p. 17, note.
Habitat. The Islands of St. Thomas and St. Croix.

"This bird," says Mr. Edward Newton, speaking of the Humming-Birds of St. Thomas and St. Croix, "breeds from the end of March to the end of June. It is no easy matter to find its nest; for on approaching within two or three yards of where it is, the bird, if it is on, is sure to fly at you, and then retreating remains suspended a few seconds just above your head, when it darts off and perches on some dead twig, most likely on the very tree which holds its nest. It does not stay here long, but takes short flights into the air, returning to the same place and, when there, showing its impatience by a continual flirting, or rather twitching of its wings. If you then retire, keeping your eye on the bird, it will presently dart straight on to its nest, leaving it, however, at the least movement on your part. This species is not particular as to the tree on which it builds, as I have found nests on the Silk-cotton, Mango, Manchioneel, Avocado-Pear (*Laurus persea*, Linn.). They are placed on a horizontal branch, from half an inch to two inches thick, and are composed of cotton or the down of a species of *Cactus*, studded on the outside with white Lichen or shreds of bark, the whole structure measuring nearly two inches across, and built at the height of from about five to fifteen feet from the ground, sometimes concealed by leaves, at others on an almost naked bough."—*Ibis*, vol. i. p. 140.

Mr. Newton informs me that the yellow of the base of the bill and gape of this bird shows rather conspicuously.

95. EULAMPIS LONGIROSTRIS, *Gould*.

In its size, general plumage, and style of colouring, this bird is very similar to the *E. chlorolæmus*; but the much greater length and curvature of its bill will, I am sure, satisfy the most sceptical that it is quite distinct. I possess two examples of this, both of which are unfortunately in a very bad state of plumage. One of these was presented to me by my valued friend Sigismund Rucker, Esq., the other I obtained on the continent; I could gain no information whatever as to its native locality. The average length of the bill in *E. chlorolæmus* is three-quarters of an inch, while that of *E. longirostris* is nearly an inch and a quarter.

Habitat. Unknown.

There is scarcely a more isolated form in the family of *Trochilidæ* than that for which the generic name of *Lafresnaya* was proposed by Bonaparte in honour of the venerable Baron de Lafresnaye, and it gives me great pleasure to assist in perpetuating the name of a French nobleman, lately deceased, who devoted the leisure hours of a long life to the pleasing study of natural history.

Strictly confined to the Andes, one of the species is quite equatorial, the others fly several degrees further north. The males are very boldly coloured, the brilliant green of their throats and flanks being beautifully relieved by the velvety black of the abdomen. The females have none of these contrasted colours, their entire under surface being spangled with green on a white or a buff ground. The species known are very much alike except in the colouring and markings of the tail,—one of them having the four outer feathers white tipped with purplish black, while the same feathers in another are buff tipped with bronzy brown, and the tail of the third is white tipped with greenish bronze.

Genus LAFRESNAYA, *Bonap.*

96. LAFRESNAYA FLAVICAUDATA . . . Vol. II. Pl. LXXXV.
Trochilus flavicaudatus, Fras.
——— *Lafresnayi*, Bois.
Calothorax Lafresnayi, Gray & Mitch.
Lafresnayi flavicaudatus, Bonap.
——— *flavicaudata*, Reichenb., Bonap.
Entima Lafresnayi, Cab. et Hein. Mus. Hein. Theil iii. p. 51.

Habitat. The high lands of New Granada. Common at Bogota and Popayan; and probably in the northern parts of Ecuador.

97. LAFRESNAYA GAYI . . . . . . Vol. II. Pl. LXXXVI.
Trochilus Gayi, Bourc. et Muls.
Calothorax Gayi, Gray & Mitch.

Lafresnaya gayi, Bonap.
*Entimia Gayi, Cab. et Hein. Mus. Hein. Theil iii. p. 51.
Habitat. Ecuador and Peru.

98. LAFRESNAYA SAULÆ.
* Trochilus Saulæ, Bourc. Rev. Zool. 1846, p. 309.
* Calothorax Saulii, Gray & Mitch. Gen. of Birds, vol. i. p. 110, Calothorax, sp. 3.
* Lafresnaya Saulæ, Bonap. Consp. Gen. Av. tom. i. p. 68, Lafresnaya, sp. 3.
* ———— Saul, Reichenb. Aufz. der Col. p. 11.

Habitat. Unknown: supposed to be Popayan.

Since writing my account of *Lafresnaya Gayi* I have received many additional examples, all of which had white tails tipped with purplish black; but I possess fully adult examples of a white-tailed bird named *Saulæ*, by M. Bourcier, in which the tippings are bronzy green. My specimens were brought by Delattre; but from what locality, is unknown. The difference mentioned seems to warrant the belief that the bird is distinct; and I therefore give it a place in this synopsis, notwithstanding the opinion to the contrary expressed in my account of *L. Gayi*.

Those who have not closely studied the Humming-Birds have but little idea how diversified are their forms; the birds next to be considered are unlike all the other members of the family. The species are short, thick-set birds, with a very peculiar style of plumage, have their crowns plated with metal-like feathers, and bills as straight and sharp as needles; and woe to any bird, I should say, which gave offence to the members of this genus.

I am exceedingly puzzled with respect to the species of this form; that is, I am at a loss to determine whether they are two, three, four, or five in number. First, with regard to *Johannæ*, whose under-surface is black, and frontal mark violet-blue; I have always regarded this colouring as indicative of the adult, but I am in doubt whether the skins which frequently accompany them from Bogota, and which assimilate in size and form, but differ in having a green frontlet and a dull-green upper and under surface, are the females or young males of this bird, or if they be distinct. Of the *Ludoviciæ*, which comes from Bogota, I have many examples, all of which are very uniform in size and style of colouring. From Quito I have another bird assimilating to the *Ludoviciæ* most closely in colouring, but which is about a fifth larger in all its admeasurements. Accompanying the specimens from this latter locality is one without any frontal mark whatever; in other respects it is precisely like the rest, and, I am sure, is a fully adult bird. Is this the female of the Quitan birds, or a distinct species? I have never seen examples in this state of plumage among the numerous specimens sent from Bogota. I think I shall be right in regarding the Ecuadorian bird as distinct, and I therefore propose for it the name that of *rectirostris*.

Genus DORYFERA, *Gould.*
(Δόρυ, hasta, et φέρω, fero; Lance-bill.)

Generic characters.

Male.—*Bill* long, basal half straight, apical half inclined upwards and pointed; *wings* of moderate size; *tail* rounded, the feathers broad and rigid; *tarsi* partly clothed; *hind toe* and *nail* as long as the middle toe and nail; *forehead* luminous; *plumage* adpressed.

Female.—I believe the female is destitute of the forehead mark; but this is uncertain.

99. DORYFERA JOHANNÆ . . . . . Vol. II. Pl. LXXXVII.

Trochilus Johannæ, Bourc.
Mellisuga Johannæ, Gray & Mitch.
Trochilus (Doryfera) violifrons, Gould.
Dorifera Johannæ, Bonap.
* *Hemistephania Johannæ,* Reichenb. Aufz. der Col. p. 9.
* *Helianthea Johanna,* Reichenb. Troch. Enum. p. 6, pl. 731. figs. 4675-76.
* *Doryphora Johanna,* Cab. et Hein. Mus. Hein. Theil iii. p. 78, note.

Habitat. New Granada.

100. DORYFERA LUDOVICIÆ . . . . Vol. II. Pl. LXXXVIII.

Trochilus Ludoviciæ, Boure. et Muls.
Mellisuga Ludoviciæ, Gray & Mitch.
Dorifera ludoviciæ, Bonap.
* *Hemistephania Ludoviciæ,* Reichenb. Aufz. der Col. p. 9.
* *Helianthea Ludoviciæ,* Reichenb. Troch. Enum. p. 6, pl. 731. figs. 4673-74.
Doryphora Ludoviciæ, Cab. et Hein. Mus. Hein. Theil iii. p. 77.

Habitat. New Granada.

101. DORYFERA RECTIROSTRIS, *Gould.*

Bill and feet black; tarsi clothed with brown feathers; forehead brilliant glittering green; crown and back of the neck reddish bronze, passing into dull green on the back; upper tail-coverts washed with blue; tail black, tipped with greyish-brown, largely on the external feathers, slightly on the middle ones; under surface olive; under tail-coverts grey; wings purplish brown.

Total length 5 inches; bill 1$\frac{1}{4}$; wing 2$\frac{3}{4}$; tail 1$\frac{1}{2}$.

Habitat. Ecuador.

How remarkable it is that development and even beauty should be bestowed upon the under tail-coverts of a bird! yet this is often found to be the case: the Marabou Stork may be cited as an instance in point among the larger birds, and the genera *Eriocnemis, Erythronota,* &c. among the Trochilidæ. In no group, however, is this feature so conspicuously marked as in the members of the succeeding genus *Chalybura*; there it is carried to its maximum and is rendered so much the more apparent from the striking contrast of the snow-white plumed under tail-coverts with the dark or black

Genus HELIODOXA, *Gould.*
("Ηλιος, sol, et δόξα, gloria.)

Generic characters.

Male.—*Bill* longer than the head, straight and cylindrical; *wings* long and pointed; *tail* ample and forked; *tarsi* clothed; *feet* small; *hind toe* shorter than the middle one; *nails* feeble; centre of the throat blue, surrounded by brilliant green.

Female.—Unadorned.

108. HELIODOXA JACULA, *Gould* . . . . Vol. II. Pl. XCIV.
   Leadbeatera jacula, Bonap., Reichenb.
   * *Coeligena jacula*, Reichenb. Troch. Enum. p. 4, pl. 688. fig. 4522.
   * *Heliodoxa jacula*, Cab. et Hein. Mus. Hein. Theil iii. p. 22.
   Habitat. New Granada.

109. HELIODOXA JAMESONI . . . . . . Vol. II. Pl. XCV.
   Trochilus Jamesoni, Bourc.
   Leadbeatera Jamesoni, Bonap., Reichenb.
   Coeligena Jamesoni, Reichenb.
   Heliodoxa Jamesoni, Sclat., Cab.
   Habitat. Ecuador.

Hitherto I have entertained the opinion that the *jacula* and *Leadbeateri* were of the same form; but upon further consideration I now believe them to be distinct; and as the former is the type of my genus *Heliodoxa*, I retain that of *Leadbeatera* for the other.

Genus LEADBEATERA, *Bonap.*

Of this form I possess three very distinct birds, which might be considered by some persons as one and the same, but in this opinion I cannot agree: the *Otero* from Bolivia, and the *Leadbeateri* are too unlike to be considered otherwise than as separate species; while the third, which is from Venezuela, is allied to the Bolivian bird rather than to that from New Granada.

110. LEADBEATERA OTERO.
   Heliodoxa Otero . . . . Vol. II. Pl. XCVI.
   Trochilus Otero, Tschudi.
   Leadbeatera otero, Bonap., Reichenb.
   * *Colligena Otero*, Reichenb. Troch. Enum. p. 3, pl. 689. figs. 4523–24.
   * *Heliodoxa Otero*, Cab. et Hein. Mus. Hein. Theil iii. p. 22, note.
   * *Leadbeatera sagitta*, Reichenb. Aufz. der Col. p. 7.
   * *Coeligena sagitta*, Reichenb. Ib. p. 23; Id. Troch. Enum. p. 4, pl. 689. fig. 4525, and pl. 690. figs. 4527–28.
   * *Heliodoxa sagitta*, Cab. et Hein. Mus. Hein. Theil iii. p. 22.
   Habitat. Peru and Bolivia.

111. LEADBEATERA SPLENDENS, *Gould.*
   Centre of the crown brilliant blue, bordered on each side with jet-

black; upper surface bronzy green; wings purplish brown; two centre tail-feathers bronzy, the remainder black; under surface glittering green; under tail-coverts olive-grey; bill black; feet dark brown.

Total length, 5½ inches; bill 1 1/16; wing 2⅜; tail 2¼; tarsi ¼.

Habitat. Venezuela.

This species is very nearly allied to the *Leadbeatera Otero*, but it differs in having a straighter and shorter bill, and in the green tint of the under surface.

112. LEADBEATERA GRATA.

Heliodoxa Leadbeateri . . . . Vol. II. Pl. XCVII.

Trochilus Leadbeateri, Bourc.
Leadbeatera grata, Bonap.
Mellisuga Leadbeateri, Gray & Mitch.
Heliodoxa Leadbeateri, Sclat.

Habitat. The hilly parts of New Granada.

It matters not where we place the single species of the genus *Aïthurus* (*Trochilus polytmus*, in the body of the work), since it offers no direct alliance to any one group. It is perhaps the most singular and most aberrant of Humming-Birds: for it departs from all the rest in the form of its wings, the second feather being the longest, while in all the others the first exceeds the rest in length; how different also are its other characters! for instance, the tail is not forked in the usual way, the second feather being lengthened into flowing plumes, which apparently tend more to add to its graceful appearance than to facilitate its aërial evolutions. The young males do not possess this peculiarly formed tail; and the females are so unlike both, that we should not have even suspected their alliance, had we not positive evidence of it. This very isolated form is a native of Jamaica, and there alone is it found. That so large a bird and so very marked a form should be confined to such a limited area is very surprising.

Genus AÏTHURUS, *Cab.*

113. AÏTHURUS POLYTMUS.

Trochilus polytmus . . . . Vol. II. Pl. XCVIII.

Mellisuga Jamaicensis atricapilla, cauda bifurca, Briss.
Mellivora avis maxima, Sloane.
Trochilus polytmus, Linn. et auct.
Ornismya cephalatra, Less.
* *Cynanthus polytmus*, Jard. Nat. Lib. Humming-Birds, vol. ii. p. 145.
* *Polytmus cephalatra*, Bonap. Consp. Gen. Av. tom. i. p. 72, *Polytmus*, sp. 1.
* ————— Ib. sp. 2.; *Trochilus stellatus*, "Gosse," young male?
* *Trochilus Maria*, Hill, Ann. and Mag. Nat. Hist. ser. 2. vol. iii. p. 258, 1849; Gosse, Ill. Birds of Jamaica, pl. 22.

*_Polytmus viridans_, Reichenb. Aufz. der Col. p 11; Id. Troch. Enum. p. 9, pl. 799. figs. 4858–60.
Aithurus polytmus, Cab. et Hein. Mus. Hein. Theil iii. p. 50.
Habitat. Jamaica.

It will be seen that I have placed the _stellatus_ of Gosse as a synonym of _Polytmus_; at the same time it is only justice to state that I have never seen a second specimen in a similar state of plumage, and it may be another species. I make this remark with Mr. Gosse's type specimen before me, it having been kindly presented to me by that gentleman.

Genus THALURANIA, _Gould_.

(Θάλλω, vireo, et οὐράνιος, cœlestis.)

If all genera were as well defined as that of _Thalurania_, the ornithologist would be far less perplexed than he frequently is with regard to the position of the species of which they are composed. All the members of this extensive group are characterized by great elegance of contour, the bill, wings, and tail being well proportioned, and in harmony with the size of the body; green and blue are the prevailing hues of the under surface, while the crown and throat, and sometimes the shoulders, are ornamented with blue. The females are less elegant in form, and not so beautifully attired, all those parts which are green and blue in the males being, in every instance I believe, of a dull grey.

The extent of country ranged over by the members of this group is very great: one, and one only, has been found to the north of Panama; the remainder inhabit all the countries southward to the latitude of Rio de Janeiro.

114. THALURANIA GLAUCOPIS . . . . . Vol. II. Pl. XCIX.

Trochilus glaucopis, Gmel., Pr. Max., Vieill., Jard.
Mellisuga Brasiliensis, cauda bifurca, Briss.?
Ornismya glaucopis, Less.
Polytmus glaucopis, Gray & Mitch.
Trochilus frontalis, Lath.
Thalurania glaucopis, Bonap., Reichenb.
Coeligena glaucopis, Reichenb.
*_Cynanthus glaucopis_, Jard. Nat. Lib. Humming-Birds, vol. II. p. 147.
*_Glaucopis frontalis_, Burm. Th. Bras. tom. ii. p. 333.
*_Thalurania glaucopis_, Cab. et Hein. Mus. Hein. Theil iiL p. 29.
Habitat. South-eastern Brazil.

115. THALURANIA WATERTONI . . . . . . Vol. II. Pl. C.

Trochilus Watertoni, Lodd., Bourc.
Polytmus Watertoni, Grey & Mitch.
Thalurania Watertoni, Bonap.
——— _Watertoni_, Reichenb.
Coeligena Watertoni, Reichenb.
Habitat. British Guiana; and Northern Brazil?

116. THALURANIA FURCATA . . . . Vol. II. Pl. CI.

Mellisuga Jamaicensis, violacea, cauda bifurca, Briss.
Trochilus furcatus, Gmel., Lath., Shaw, Vieill., Steph., Jard.
Ornismya furcata, Less.
Polytmus furcatus, Gray & Mitch.
Thalurania furcata, Gould, Bonap.
——————*furcatus*, Bonap.
* *Cynanthus furcatus*, Jard. Nat. Lib. Humming-Birds, vol. ii. p. 148.
* *Thalurania Gyrinna*, Reichenb. Aufz. der Col. p. 7.
* *Coeligena Gyrinna*, Reichenb. Troch. Enum. p. 3, pl. 682. figs. 4500–1.
* *Thalurania furcata*, Cab. et Hein. Mus. Hein. Theil iii. p. 24.
* ——————*furcata*, Reichenb. Aufz. der Col. p. 7.
* *Coeligena furcata*, Reichenb. Troch. Enum. p. 3, pl. 682. figs. 4498–99.
* *Trochilus furcatus*, Burm. Th. Bras. tom. ii. p. 335.

Habitat. Cayenne and Guiana.

117. THALURANIA FURCATOIDES, *Gould.*

Thalurania furcatoides, Gould, in text to *T. furcata.*

Habitat. Para and the lower part of the Amazon.

This bird is very like *T. furcata*, but is of smaller size, has a much less-forked tail, and the breast ultramarine-blue instead of purplish-blue. I have at this moment seven specimens before me, all of which are alike, and readily distinguishable from the Cayenne bird.

118. THALURANIA FORFICATA, *Cab.*

* *Thalurania forficata*, Cab. et Hein. In Mus. Hein. Theil iii. p. 24.

Habitat. Supposed to be the neighbourhood of Para.

Through the kindness of Dr. Peters, the Director of the Royal Museum of Berlin, I have had their specimen of *T. forficata* sent to me for comparison. It is certainly distinct from any of the species contained in the collections of this country: in size it is nearly the same as *T. furcata*; but its bill is rather shorter, its tail longer, more deeply forked, and of a purplish-black colour instead of steely black; it differs from that bird also in having the blue colouring more extended down the back, approaching to that of *Watertoni*; the green of the throat is circumscribed and truncate below, as in *furcata*; and the crown of the head is black, but near the centre is a single small blue feather: I think it likely that this is accidental, as the bird appears to be fully adult: the under tail-coverts are black.

119. THALURANIA REFULGENS, *Gould* . . . Vol. II. Pl. CII.

Thalurania refulgens, Gould, Bonap., Reichenb.
Coeligena refulgens, Reichenb.

Habitat. Unknown; supposed to be Venezuela.

120. THALURANIA TSCHUDII, *Gould* . . . . Vol. II. Pl. CIII.
Trochilus furcatus, Tschudi.
Habitat. Ucayali and the eastern part of Ecuador.

121. THALURANIA NIGROFASCIATA, *Gould*. . Vol. II. Pl. CIV.
Trochilus (— ?) *nigrofasciatus*, Gould.
Thalurgnia nigrofasciata, Gould, Reichenb., Cab.
——————— *nigrofasciatus*, Bonap.
Polytmus nigrofasciatus, Gray & Mitch.
Thalurania nigrofasciata, Bonap.
Coeligena nigrofasciata, Reichenb.
Thalurania viridipectus, Gould, Bonap., Reichenb.
Coeligena viridipectus, Reichenb.
Habitat. Upper Amazon and Rio Napo.

122. THALURANIA VENUSTA, *Gould* . . . . Vol. II. Pl. CV.
Trochilus (*Thalurania*) *venusta*, Gould.
——— (—————) *puella*, Gould.
Thalurania venusta, Gould, Bonap., Reichenb.
————— *puella*, Bonap., Reichenb.
* *Coeligena venusta*, Reichenb., Troch. Enum. p. 3, pl. 683. figs. 4504–5.
* ————— *puella*, Reichenb. Ib. p. 3.
Habitat. Chiriqui in Costa Rica.

123. THALURANIA COLUMBICA . . . . . . Vol. II. Pl. CVI.
Ornismya Colombica, Bourc. et Muls.
Polytmus Colombicus, Gray & Mitch.
Thalurania Columbiana, Gould.
——————— *colombica*, Bonap.
——————— *columbica*, Reichenb.
* ——————— *columbica*, Cab. et Heln. Mus. Heln. Theil iii. p. 24.
Habitat. The hilly parts of New Granada.

124. THALURANIA VERTICEPS, *Gould* . . . Vol. II. Pl. CVII.
Trochilus (*Thalurania*) *verticeps*, Gould.
Thalurania verticeps, Gould, Bonap.
Thalurania Lydia, Reichenb.
Riccordia verticeps, Reichenb.
* *Chlorestes verticeps*, Reichenb. Troch. Enum. p. 4, pl. 705. fig. 4590.
Habitat. Ecuador.

125. THALURNIA FANNIÆ.
Trochilus Fannyi, Bourc. et Delatt.
Hylocharis Fannyi, Gray & Mitch.
* *Coeligena Fanny*, Reichenb. Troch. Enum. p. 3, pl. 683. figs. 4502–3.

* *Thalurania Fanny*, Bonap. Rev. et Mag. de Zool. 1854, p. 254;
Cab. et Hein. Mus. Hein. Theil iii. p. 29.

Habitat. The Andes of Quindios.

I have a specimen of this bird, collected by Warszewicz on the
Cordillera of Quindios, which is precisely the same with the type
specimen of the *T. Fannyi* of MM. Bourcier and Delattre. It
differs from my *T. verticeps* in being considerably smaller and in
having the abdomen purple-blue in lieu of cold prussian-blue.

126. THALURANIA ERIPHYLE . . . . Vol. II. Pl. CVIII.

Ornismya Eriphile, Less.
Polytmus Eriphile, Gray & Mitch.
Thalurania eryphila, Bonap.
——————— *Eryphile*, Reichenb.
——————— *eriphile*, Bonap., Gould.
Ornismya meriphile, Less. in err.
* *Cœligena Eryphile*, Reichenb. Troch. Enum. p. 3, pl. 582. figs.
4507-8.
* *Glaucopis eriphile*, Burm. Th. Bras. tom. ii. p. 334.
* *Trochilus excisus*, Licht. in Mus. of Berlin.
* *Thalurania eryphile*, Cab. Mus. Hein. Theil iii. p. 29.

Habitat. Eastern Brazil.

127. THALURANIA? WAGLERI . . . . . . Vol. II. Pl. CIX.

Ornismya Waglerii, Less.
Trochilus Waglerii, Jard.
Cynanthus Waglerii, Jard.
Hylocharis Wagleri, Gray & Mitch.
Thalurania wagleri, Bonap., Reichenb.
* *Trochilus bicolor*, Vieill. Ois. dor., tom. l. p. 75, pl. 36.
* *Cœligena Wagleri*, Reichenb. Troch. Enum. p. 3, pl. 702. figs.
4576-77.
* *Thalurania Wagleri*, Cab. et Hein. Mus. Hein. Theil iii. p. 24.

Habitat. Brazil.

The three species constituting my genus *Panoplites* very closely
resemble each other in size, in structure, and in the markings of
their tails, but are very dissimilar in the colouring of their bodies.
They are all inhabitants of the Ecuadorian Andes, and one of them
(*P. flavescens*) extends its range as far north as Bogota. The
most remarkable species of this form is the *P. Jardinei*, whose glittering upper surface is wonderfully brilliant.

Genus PANOPLITES, *Gould.*

(Πανοπλίτης, omnino armatus.)

Generic characters.

Male.—Bill strong, and a trifle longer than the head; *body* stout
and thick-set; *wings* long and pointed; *tail* moderately long and
square, the feathers broad; *tarsi* clothed and stout; *hind toe* strong,
and of the same length as the middle one; *nails* short.

Female.—Very similar to the male in plumage.

G 2

128. PANOPLITES JARDINEI . . . . Vol. II. Pl. CX.
Trochilus Jardini, Bourc.
Florisuga jardinii, Bonap., Reichenb.
**Panoplites Jardinei*, Cab. et Hein. Mus. Hein. Theil iii. p. 75, note.
Habitat. Nanegal and other warm parts of Ecuador.

129. PANOPLITES FLAVESCENS . . . . . . Vol. II. Pl. CXI.
Trochilus flavescens, Lodd.
Ornismia paradisea, Boiss.
Mellisuga flavescens, Gray & Mitch.
Amazilius flavescens, Bonap.
Clytolæma flavescens, Bonap.
**Boissonneaua flavescens*, Reichenb. Aufz. derCol. p. 11; Id. Troch. Enum. p. 8, pl. 787. figs. 4830–31.
**Panoplites flavescens*, Cab. et Hein. Mus. Hein. Theil iii. p. 74.
**Trochilus Lichtensteinii*, Saucer. in Mus. of Berlin.
Habitat. Ecuador and New Granada.

130. PANOPLITES MATTHEWSI . . . . . Vol. II. Pl. CXII.
Trochilus Matthewsi, Lodd., Bourc.
Mellisuga Matthewsi, Gray & Mitch.
Clytolæma matthewsi, Bonap.
**Heliodoxa Matthewsii*, Reichenb. Aufz. der Col. p. 9.
**Boissonneaua Matthewsii*, Reichenb. Troch. Enum. p. 8, pl. 787. figs. 4832–33.
**Panoplites Matthewsi*,Cab.et Hein.Mus.Hein.Theil iii. p.74,note.
Habitat. Ecuador.

Somewhat allied to the *Panoplitæ*, are the members of the

Genus FLORISUGA, Bonap.,

all of which are remarkable for their large fan-shaped tails, and for having all the feathers of this organ white. The females are less strikingly coloured. One of the species, *F. mellivora*, enjoys a most extensive range; for it inhabits alike the low lands of Northern Brazil, Cayenne, Guiana, Trinidad, Venezuela, the temperate regions of New Granada, and Central America; the other two are confined to more limited areas. The *F. mellivora* and *F. atra* are among the oldest-known and the commonest of the Humming-Birds, there being no collection of any extent without examples of them.

131. FLORISUGA MELLIVORA . . . . . . Vol. II. Pl. CXIII.
Trochilus mellivorus, Linn. et auct.
Mellisuga Surinamensis torquata, Briss.
———— *Surinamensis*, Steph.
Ornismya mellivora, Less.
Topaza mellivora, Gray & Mitch.
Florisuga mellivora, Bonap.
Trochilus fimbriatus, Linn., Lath.

Mellisuga Cayenensis gutture nævio, Briss.
Topaza fimbriata, Gray & Mitch.
Lampornis mellivora, Jard. Nat. Lib. Humming Birds, vol. ii. p. 155.
Florisuga mellivora, Reichenb. Aufz. der Col. p. 14; Id. Troch. Enum. p. 12; Cab. et Hein. Mus. Hein. Theil iii. p. 29.

Habitat. Brazil, Trinidad, New Granada, Bogota, and Guatemala.

132. FLORISUGA FLABELLIFERA, *Gould* . . Vol. II. Pl. CXIV.
Trochilus (——— ?) *flabelliferus*, Gould.
Florisuga flabellifera, Bonap.
Florisuga flabellifera, Reichenb. Aufz. der Col. p. 14; Id. Troch. Enum. p. 12; Cab. et Hein. Mus. Hein. p. 29, note.
Topaza flabellifera, Gray & Mitch. Gen. of Birds, vol. i. p. 110.
Topaza, sp. 8.

Habitat. The Island of Tobago, and perhaps elsewhere.

"I am not able," says Mr. Kirk, "to decide as to this bird being a native of Tobago. It is only to be met with at certain seasons; but whether it leaves the island, or retires to the interior, I am not at present prepared to say. It is seldom to be found in open sunshine: the mornings and evenings are its principal times for feeding, and its evolutions then are truly pleasing,—at one instant suspended immovable to the eye (although alternately showing the purest white and green), at the very top of our tallest bamboo, guava, or other tree, and at the next moment at the root, with two or three zigzags right and left, up and down, dipping either into the river or snapping a fly from the surface, and then disappearing. I think it probable that this bird feeds more upon winged insects than most of the others, which may account for its being seen so early in the calm mornings, retiring generally into the thick wild plantain bushes as soon as the sun begins to spread his rays upon them, and appearing again in the evening when he is going down, or when his rays cease to act upon their spot of pleasure. A female shot on the 19th of April contained an egg almost perfect."—*Horæ Zoologicæ*, by Sir W. Jardine, Dt., in *Ann. and Mag. Nat. Hist.* vol. xx. p. 373.

133. FLORISUGA ATRA . . . . Vol. II. Pl. CXV.
Trochilus ater, Pr. Max.
——— *atratus*, Licht.
——— *niger*, Swains.
——— *fuscus*, Vieill.
Ornismya lugubris, Less.
Colibri leucopygius, Spix.
Mellisuga ater, Steph.
Topaza atra, Gray & Mitch.
Florisuga atra, Bonap.
Lampornis niger, Jard. Nat. Lib. Humming Birds, vol. ii. p. 156.
Florisuga fusca, Reichenb. Aufz. der Col. p. 14; Id. Troch. Enum. p. 12; Cab. et Hein. Mus. Hein. Theil iii. p. 29.

Habitat. Eastern Brazil.

That all the Humming-Birds are not yet discovered we may very reasonably conjecture, and we may ask what will be our next novelty in this group of birds. This remark has suggested itself upon finding next in succession the singular little *Microchæra albocoronata.* Although America has been discovered for more than 300 years, and collectors have been employed for the last fifty in searching for its treasures of various kinds, we had no knowledge of the existence of this bird until 1852, when Dr. J. K. Merritt shot three examples in the district of Belen in New Granada.

Genus MICROCHÆRA, *Gould.*
(μικρὸὶ, parvus, et χήρα, vidua.)

Generic characters.

Male.—Bill about the same length as the head, and straight; *wings* moderate; *tail* rather short and square; *tarsi* clothed; *feet* small; *claws* diminutive.

134. MICROCHÆRA ALBOCORONATA, *Gould.* . Vol. II. Pl. CXVI.

Mellisuga albo-coronata, Lawr.

Habitat. The district of Belen in New Granada.

A considerable hiatus here occurs, which cannot at present be filled up, and we come to the elegant frill-necked Coquettes, the *Lophornithes*; and with these I commence the third volume.

They are among the most beautiful of the smaller members of the Trochilidæ, and are distinguished by the possession of lengthened ornamental plumes springing from the sides of the neck, which generally have a spangle of metallic lustre at the tip; they are further ornamented with beautiful lengthened crests, which are developed to a greater extent in some species than in others; in those in which the neck plumes are the longest, the crests are least so, and *vice versâ.* They are spread over a great part of America, from Mexico along the Andes to Bolivia; they also occur in Brazil, the Guianas, and the Island of Trinidad.

Genus LOPHORNIS, *Less.*

135. LOPHORNIS ORNATUS . . . . . Vol. III. Pl. CXVII.

Trochilus ornatus, Gmel., Lath., Shaw.
Ornismya ornata, Less.
Mellisuga ornata, Gray & Mitch.
Lophornis auratus, Bonap.
———— *ornatus,* Bonap.
———— *ornata,* Less., Gray, Reichenb.

Habitat. Northern Brazil, Guiana, and Trinidad.

Mr. W. Tucker informs me that in Trinidad this species "frequents the pastures and open places, and visits the flowers of all the small shrubs, but is particularly fond of those of the Ipecacuanha plant, and that it is very pugnacious, erecting its crest, throwing out its whiskers, and attacking every Humming-Bird that may pass within its range of vision."

136. LOPHORNIS GOULDI . . . . . Vol. III. Pl. CXVIII.

Ornismya Gouldii, Less.
Trochilus Gouldii, Jard.
Lophornis Gouldii, Less.
Mellisuga Gouldi, Gray & Mitch.
Lophornis gouldi, Bonap.
Bellatrix Gouldii, Reichenb.

Habitat. Northern Brazil.

137. LOPHORNIS MAGNIFICUS . . . . Vol. III. Pl. CXIX.

Trochilus magnificus, Vieill., Temm., Jard., Pr. Max. zu Wied.
————— *decorus*, Licht.
Colibri helios, Spix.
Ornismya magnifica, Less.
————— *strumaria*, Less.
Lophornis strumaria, Less.
Mellisuga magnifica, Gray & Mitch.
Lophornis magnificus, Bonap.
Bellatrix magnifica, Reichenb.
*Ornismya strumaria, Dev. Rev. et Mag. de Zool. 1852, p. 215.

Habitat. South-eastern Brazil.

Mr. Deville states that this bird is found for a portion of the year in different Brazilian districts, and is so bold that the sight of man creates no alarm. Its food chiefly consists of small insects, which it seizes on the wing, precipitating itself from the extremity of a dead bough, upon which it often passes entire hours in the same position; when it has chosen a branch it rarely proceeds far from it, and always returns to it. It is very common in the environs of Rio de Janeiro.

138. LOPHORNIS REGULUS, *Gould* . . . Vol. III. Pl. CXX.

Trochilus (Lophornis) regulus, Gould.
Mellisuga regulus, Gray & Mitch.
Lophornis regulus, Bonap.

Habitat. Cochabamba in Bolivia.

I possess a bird of this genus from Peru, with a more truncate form of crest than that of *L. Regulus*, the fine feathers of which are rather largely tipped with spangles of dark green. This may probably prove to be, and I believe is, really distinct; I have consequently proposed for it the specific name of *lophotes*. In size and colouring it very closely resembles the *L. Regulus*, with the exceptional difference in the form of the crest.

139. LOPHORNIS LOPHOTES, *Gould*.

Habitat. Peru.

140. LOPHORNIS DELATTREI, *Less.* . . . Vol. III. Pl. CXXI.
Habitat. New Granada.

141. LOPHORNIS REGINÆ, *Gould* . . . Vol. III. Pl. CXXII.
Lophornis Reginæ, Gould.
Mellisuga reginæ, Gray & Mitch.
Lophornis reginæ, Bonap.
Bellatrix Reginæ, Reichenb.
Habitat. New Granada.

Mr. Fraser, who killed an example at Zamora, in Ecuador, states that the irides of this species are black, and its mandibles reddish flesh-colour, with a black tip; he adds that it was feeding from a large Guarumba tree.

142. LOPHORNIS HELENÆ . . . . . . Vol. III. Pl. CXXIII.
Orniemya Helenæ, Delatt.
Mellisuga Helenæ, Gray & Mitch.
Lophornis helenæ, Bonap.
——— *helena*, Bonap.
——— *Helenæ*, Reichenb.
Habitat. Guatemala and Southern Mexico.

Mr. Salvin states that this species is not uncommon in the vicinity of Coban, and that its cry "is peculiarly shrill and unlike that of any other species I know; hence its presence may be noticed if only the cry of a passing bird be heard. It feeds among the *Salviæ* that so abound in the mountain-hollows about Coban; and it is said also to show a partiality for the flowers of the *Tasisco* when that tree is in full bloom, in the month of December. In the month of November females of this species are very rare. Of the specimens I collected, there was only one female to seventeen males.

"In the Indian language of Coban, *Lophornis Helenæ* has, besides the name 'Tzunnun,' which is applied to all the small Humming-Birds, the additional name of '*Achshukub*.' The Spanish name is '*El Gorrion Cachudo*'—the Horned Humming-Bird."—*Ibis*, vol. ii. p. 268.

Although I have placed all the species known by the trivial name of Coquettes in the genus *Lophornis*, the *L. chalybeus* and *L. Verreauxi* have been separated by M. Cabanis into a distinct genus, under the name of *Polemistria*.

Genus POLEMISTRIA, *Cab.*

These birds, as will be seen on reference to the plates on which they are represented, vary considerably from all the true *Lophornithes*; the feathers of the neck-frill are very different, and the tail is much longer and more rounded. I shall not be surprised if another species of this peculiar form should be discovered; for I have in my possession the skin of a female from Bogota, which I am inclined to think is the female of an unknown species.

143. POLEMISTRIA CHALYBEA.
Lophornis chalybeus . . . . . . Vol. III. Pl. CXXIV.
Trochilus chalybeus, Vieill., Temm., Jard.
——— *festivus*, Licht.
Ornismya Vieillottii, Less.
Mellisuga chalybea, Gray & Mitch.
Lophornis chalybeus, Bonap.
Ornismya Audenetii, Less.
Trochilus Audenetii, Jard.
Colibri mystax, Spix.
Mellisuga Audenetii, Gray & Mitch.
Lophornis Audeneti, Bonap.
Habitat. Brazil.

144. POLEMISTRIA VERREAUXI.
Lophornis Verreauxi . . . . . . Vol. III. Pl. CXXV.
Trochilus Verreauxi, Bourc.
Lophornis Verreauxi, Reichenb.
Bellatrix verreauxi, Bonap.
Habitat. Peru.

I shall now proceed to the single species of the genus *Discura*. The band which crosses the lower part of the back allies this bird to the *Lophornithes* on the one hand, and to *Prymnacantha* and the *Gouldiæ* on the other.

Genus DISCURA, *Bonap.*
145. DISCURA LONGICAUDA . . . . . Vol. III. Pl. CXXVI.
Trochilus longicaudus, Gmel.
Mellisuga longicauda, Gray & Mitch.
Discosura longicauda, Bonap.
Discura longicauda, Bonap., Reichenb.
Trochilus platurus, Lath., Vieill., Drapiez, Pr. Max. zu Wied.
Ornismya platura, Less.
Mellisuga platura, Steph.
Trochilus (Ocreatus) ligonicaudus, Gould.
Discosura ligonicauda, Bonap.
Discura platura, Reichenb.
Habitat. Cayenne, Brazil, and Demerara.

Gouldia, Popelairia, Gouldomyia and *Prymnacantha*, are all generic terms proposed for the four species I have called by the trivial name of Thorn-tail. Of these Bonaparte's name of *Gouldia*, having the priority, has been adopted by me in the body of this work ; but as the first species G. *Popelairi* differs from the others in possessing a most singular and elegant crest terminating in two hair-like feathers, I propose to adopt M. Cabanis's classical name of *Prymnacantha* for this species, and to retain *Gouldia* for the others.

Genus PRYMNACANTHA, *Cab.*

146. PRYMNACANTHA POPELAIREI.
Gouldia Popelairi . . . . Vol. III. Pl. CXXVII.
Trochilus Popelairii, Dubus.
Mellisuga Popelairii, Gray & Mitch.
Gouldia popelairi, Bonap.
Habitat. New Granada.

Genus GOULDIA, *Bonap.*

147. GOULDIA LANGSDORFFI . . . Vol. III. Pl. CXXVIII.
Trochilus Langsdorffi, Vieill., Temm., Valenc., Less.
Ornismya Langsdorffi, Less.
Colibri hirundinaceus, Spix.
Mellisuga Langsdorffi, Gray & Mitch.
Gouldia langsdorffi, Bonap.
Habitat. Brazil; and I have a single specimen from the Rio Napo.

148. GOULDIA CONVERSI . . . . . Vol. III. Pl. CXXIX.
Trochilus Conversii, Bourc.
Mellisuga Conversii, Gray & Mitch.
Gouldia conversi, Bonap.
Habitat. From Bogota along the Andes to Popayan; and Ecuador, from which latter country I have a single specimen.

149. GOULDIA LETITIÆ . . . . Vol. III. Pl. CXXX.
Trochilus Letitiæ, Bourc.
Gouldia lætitiæ, Bonap.
—— *Laetitia*, Reichenb.
Habitat. Bolivia.

Genus TROCHILUS, *Linn.*

The members of this genus as now restricted are only two in number—*T. colubris* and *T. Alexandri*. Both these birds are of moderate size and of elegant proportions. The males are decorated with richly-coloured gorgets, while the females are clothed in a sombre livery.

150. TROCHILUS COLUBRIS, *Linn.* . . Vol. III. Pl. CXXXI.
Trochilus colubris, Linn. et auct.
Mellisuga Carolinensis gutture rubro, Briss.
—— *colubris*, Steph., Gray & Mitch.
Ornismya colubris, Less.
**Cynanthus colubris*, Jard. Nat. Lib. Humming-Birds, vol. ii. p. 143.
Habitat. The eastern part of North America in summer; Mexico and Guatemala in winter, at which season it is also occasionally found in Cuba, and sometimes in Bermuda.

I have observed that specimens from Guatemala are smaller and lighter-coloured than those obtained in North America. In all probability these constitute a race which does not migrate so far north as the United States. It is probable, also, that the birds frequenting the latter country do not go further south than Mexico.

"This species," say Messrs. Sclater and Salvin, "would appear to be abundant in the winter months in Guatemala, as numerous examples were obtained by Mr. Skinner. It occurs at Acatenango, a village on the southern slope of the great Cordillera, showing that it chooses for its winter retreat the moderate climate afforded by the region lying between the elevations of 3000 and 4000 feet."—*Ibis*, vol. i. 1859, p. 129.

151. TROCHILUS ALEXANDRI, *Bourc. et Muls.*
Vol. III. Pl. CXXXII.

Trochilus Alexandri, Bourc. et Muls., Cassin.
Mellisuga Alexandri, Gray & Mitch.
Archilochus Alexandri, Reichenb.
Trochilus Alexandri, Bonap.
**Trochilus Cassini*, Bonap. MSS.
*———— *Suecicus*, in Mus. Götzian. Dresden.
**Selasphorus Alexandri*, Reich. Troch. Enum. p. 10.

Habitat. Northern Mexico and California. Found by Dr. Heermann at Sacramento city, within the limits of the United States.

Genus MELLISUGA, *Briss.*

This genus contains but a single species, unless the very minute Humming-Bird in the Loddigesian Collection should prove to be really distinct. The member or members of the genus, as the case may be, must be regarded as the very smallest of the Trochilidæ. Both sexes are destitute of luminous colouring.

152. MELLISUGA MINIMA . . . . . . Vol. III. Pl. CXXXIII.

Trochilus minimus, Linn. et auct.
Polytmus minimus variegatus, Brown.
Trochilus minutulus, Vieill.
Mellisuga Dominicensis, Briss.
Trochilus Vieilloti, Shaw.
Mellisuga Vieilloti, Steph.
Ornismya minima, Less.
Mellisuga humilis, Gosse.
Trochilus Catharinæ, Sallé.
Hylocharis nigra, Gray & Mitch.
Mellisuga minima, Bonap.
Trochilus niger, Gmel., Lath., &c.
*———— *pygmæus*, Swains. Birds of Brazil, pl. 7b.

Habitat. Jamaica and St. Domingo.

Genus CALYPTE, *Gould.*

(Καλυπτή, operta).

Generic characters.

Male.—Bill longer than the head, straight, or slightly arched;

tail rather short, the three outer feathers stiff, narrow, and slightly incurved; tarsi clothed; feet small; hind and fore toes nearly equal in length; entire head and throat luminous.

Female.—Destitute of luminous colouring.

This is strictly a Mexican genus, all the members of which are beautifully coloured, the entire head and face of the males appearing as if covered with a mask of burnished metal.

153. CALYPTE COSTÆ . . . . . . Vol. III. Pl. CXXXIV.

Ornismya Costæ, Bourc., Longuem. & Parz.
Mellisuga Costæ, Gray & Mitch.
Selasphorus Costæ, Bonap.
Atthis Costæ, Reichenb.
Trochilus Costæ, Reichenb.

Habitat. Mexico, Southern California, and Colorado Basin.

154. CALYPTE ANNÆ . . . . . . Vol. III. Pl. CXXXV.

Ornismya Anna, Less.
Trochilus Anna, Jard.
Mellisuga Anna, Gray & Mitch.
Selasphorus anna, Bonap.
Calliphlox anna, Gambel.
Atthis Anna, Reichenb.
Trochilus icterocephalus, Nuttall.
Calliphlox lamprocephalus, Licht. Cat. of Birds in Mus. of Berlin, p. 57.

Habitat. The table lands of Mexico and California.

155. CALYPTE HELENÆ . . . . . Vol. III. Pl. CXXXVI.

Orthorhynchus Helenæ, Lambeye.
——————— *Boothi*, Gundl.

Habitat. Cuba.

Genus SELASPHORUS, *Swains.*

The species of this form are characterized by the great brilliancy of the gorgets of the males. The females are destitute of these fine colours. The rounded or cuneate form of the tail in these birds at once separates them from the members of the last-mentioned genus.

156. SELASPHORUS RUFUS . . . . Vol. III. Pl. CXXXVII.

Trochilus rufus, Gmel. et auct.
——————— *collaris*, Lath.
——————— *Sitkensis*, Rathke.
Ornismya Sasin, Less.
Trochilus (Selasphorus) rufus, Swains. & Rich.
Selasphorus rufus, Audub.
Mellisuga rubra, Gray & Mitch.
Selasphorus ruber, Bonap.
Calliphlos rufa, Gambel.

Selasphorus ruber, Cab. et Hein. Mus. Hein. Theil iii. p. 56.

Habitat. Mexico. In summer it also occurs in California, and even as far north as Nootka Sound, and sometimes at Sitka.

Refer to the account of this species for my remarks respecting its not being identical with the *Trochilus ruber* of Linnæus; and also to a very interesting paper in the 'Natural History of Washington-territory,' by Drs. Cooper and Suckley.

157. SELASPHORUS SCINTILLA, *Gould*.   Vol. III. Pl. CXXXVIII.

Trochilus (Selasphorus) scintilla, Gould.

Habitat. The inner sides of the volcanic mountain Chiriqui in Costa Rica.

158. SELASPHORUS FLORESII   . . .   Vol. III. Pl. CXXXIX.

Trochilus Floresii, Lodd. MSS.

Habitat. Bolanos in Central Mexico.

159. SELASPHORUS PLATYCERCUS . . . .   Vol. III. Pl. CXL.

Trochilus platycercus, Swains.
Ornismya tricolor, Less.
——— *montana*, Less.
Mellisuga platycerca, Gray & Mitch.
Selasphorus platycercus, Bonap.
Trochilus montanus, Swains. Birds of Brazil, pl. 74.

Habitat. Guatemala, Mexico, and, according to Dr. Baird, "through Rocky Mountains to Fort Bridger, Utah."

I have observed that specimens from Guatemala are much smaller than those from the table lands of Mexico. M. Boucard found this bird at Oaxaca in Southern Mexico, and Mr. Salvin at Dueñas in Guatemala.

Genus ATTHIS, *Reichenb*.

The type of this genus is *A. Heloisæ*; and I retain the term for this very singular and beautiful bird, which in the character of its plumage and the colouring of its tail differs from every other with which I am acquainted. The plumage is extremely soft, and easily disarranged.

160. ATTHIS HELOISÆ.

Selasphorus Heloisæ . . . .   Vol. III. Pl. CXLI.

Ornismya Heloisa, Less. & Delatt.
Mellisuga Heloisa, Gray & Mitch.
Tryphæna heloisa, Bonap.

Habitat. Central America and Southern Mexico.

"Two birds were given to me," says Mr. Salvin, "by Don Vicente Constancia, who had received them from a place called Chimachoyo, near Calderas, in the Volcan de Fuego; and two other specimens I have in my collection were shot in the tierra caliente north of Coban. Hence it would appear that this, like many other species of Humming-Birds, is found in very different climates."—*Ibis*, ii. p. 266.

Numerously subdivided as the members of this family already are, I cannot pass over the beautiful *Calliope* without perceiving the necessity for a separate distinctive appellation; I therefore propose that of *Stellula*.

Genus STELLULA, *Gould*.
(dim. of *Stella*.)

Generic characters.

Male.—Bill longer than the head, straight and pointed; *wings* moderately long and sickle-shaped; *first primary* rigid; *tail* short and truncate; *feet* small; *claws* diminutive and curved.

The starry throat-mark of this bird would appear to indicate that it should be associated with the members of the next genus, *Calothorax*; but on an examination of the tail we find it to be short and truncate, and that consequently the bird is of a very different form.

At present but one species of this form has been discovered; and this ranges very far north, not only over the high lands of Mexico, but even enters California, as is shown by specimens having been sent to me from thence by Dr. Baird of Washington, and Mr. Bridges of California.

161. STELLULA CALLIOPE.

Calothorax Calliope, *Gould* . . . . . Vol. III. Pl. CXLII.
Trochilus (Calothorax) Calliope, Gould.
Calothorax calliope, Gray & Mitch., Bonap., Reichenb.

Habitat. The table lands of Northern Mexico and California.

The type of the

Genus CALOTHORAX, *G. R. Gray*,

is the well-known Mexican Star, *C. cyanopogon*, which, together with the new species discovered by M. Sallé, and named by me *C. pulchra*, are all that are known of this peculiar form. Both these birds are natives of Mexico—one inhabiting the table lands, the other the more southern and hotter districts. Both have very singularly formed tails—the outer feather being shorter than the next, and the four central ones so abbreviated as to be almost hidden by the coverts. When the males display their luminous gorgets, they must appear truly beautiful. The females possess none of this fine colouring, but, on the contrary, are very plainly attired.

162. CALOTHORAX CYANOPOGON . . . Vol. III. Pl. CXLIII.

Cynanthus Lucifer, Swains.
Ornismya cyanopogon, Less.
Calothorax lucifer, Gray & Mitch.
Trochilus cyanopogon, Jard.
———— *lucifer*, Jard.

Lucifer cyanopogon, Reichenb.
Trochilus simplex, Less.
* ———— *cyanopogon*, Swains. Birds of Brazil, pl. 77.
* ———— *coruscus*, Licht. Preis-Verz. Mexican. Thier. v. Deppe
  & Schiede (Sept. 1830) No. 34, 35.
Habitat. The table land of Northern Mexico.

163. CALOTHORAX PULCHRA, *Gould* . . Vol. III. Pl. CXLIV.
Habitat. Oaxaca in Southern Mexico.

It has always appeared to me that the well-known *Calothorax Mulsanti* and *C. Heliodori* might very properly be separated into a distinct genus; and this I have accordingly done. With these must be associated the species to which I have given the name of *C. decoratus*, and, I think, the singular little *C. micrurus*. They are all from the Andes, and are among the most diminutive of the Trochilidæ. Their structure would lead one to suspect that their wings are capable of very rapid motion, that organ being so small that it must be moved with increased rapidity to enable the bird to sustain itself when poising before a flower.

Genus ACESTRURA, *Gould*.

ἀκέστρα, acus, et οὐρά, cauda.

Generic characters.
Male.—Bill longer than the head, cylindrical, and very slightly arched; two centre tail-feathers extremely small, the two outer ones filamentous and shorter than the third; *wings* diminutive; *tarsi* clothed; *feet* small; *gorget* luminous.
Female.—Unadorned.

164. ACESTRURA MULSANTI.
Calothorax Mulsanti . . . . Vol. III. Pl. CXLV.

Ornismya Mulsanti, Boarc.
Mellisuga Mulsanti, Gray & Mitch.
Calothorax Mulsanti, Bonap.
* *Lucifer Mulsanti*, Reichenb. Troch. Enum. p. 10.
* *Trochilus filicaudus*, Licht. in Mus. Berol.
* *Chaetocercus Mulsanti*, Cab. et Hein. Mus. Hein. Theil iii. p. 60.

Habitat. The temperate regions of the Andes, from Bogota to Quito.

I observe that specimens from the former locality are smaller than those from the latter.
Mr. Fraser found this bird at Pallatanga and Puellaro in Ecuador.

165. ACESTRURA DECORATA, *Gould*.
Calothorax decoratus, *Gould* . . . . Vol. III. Pl. CXLVI.
Habitat. Uncertain, but supposed to be Antioquia, in New Granada.

166. ACESTRURA HELIODORI.
Calothorax Heliodori . . . . . . Vol. III. Pl. CXLVII.
Ornismya Heliodori, Bourc.
Mellisuga Heliodori, Gray & Mitch.
Calothorax Heliodori, Bonap.
* *Lucifer Heliodori*, Reichenb. Troch. Enum. p. 10.
* *Chaetocercus Heliodori*, Cab. et Hein. Mus. Hein. Theil iii. p. 60.
* *Calothorax bombilus*, Reichenb. MS.
Habitat. The Andes of New Granada.

167. ACESTRURA MICRURA, *Gould*.
Calothorax micrurus, *Gould* . . . . Vol. III. Pl. CXLVIII.
Habitat. Peru or Bolivia.

The members of the genus *Acestrura* naturally lead on to the *Calothorax Rosæ* and *C. Jourdani* constituting the

Genus CHÆTOCERCUS, *G. R. Gray*.

Both these birds have very singularly formed tails, as may be seen on reference to the respective plates on which they are represented.

168. CHÆTOCERCUS ROSÆ.
Calothorax Rosæ . . . . Vol. III. Pl. CXLIX.
Trochilus Rosæ, Bourc. & Muls.
Mellisuga rosæ, Gray & Mitch.
Calothorax Rosae, Reichenb.
———— *rosa*, Bonap.
* *Chaetocercus Rosae*, Cab. et Hein. Mus. Hein. Theil iii. p. 60.
Habitat. Venezuela.

169. CHÆTOCERCUS JOURDANI.
Calothorax Jourdani . . . . Vol. III. Pl. CL.
Ornismya Jourdani, Bourc.
———— *Jordani*, Bourc.
Mellisuga Jourdani, Gray & Mitch.
Calothorax jourdani, Bonap.
———— *Jourdani*, Reich.
Callothorax jourdani, Bonap.
Chætocercus Jourdani, Gray.
Habitat. The Island of Trinidad, where Mr. Tucker states that it frequents the Savannahs, but is very rare.

The bird which I have figured under the name of *Calothorax Fanny* is the type of the

Genus MYRTIS, *Reichenb.*,

which I adopt for that beautiful bird and the *C. Yarrelli*, as I consider them to be a very distinct form, and one which is especially remarkable for the structure of the tail.

170. MYRTIS FANNIÆ.
Calothorax Fanny . . . . Vol. III. Pl. CLI.
Ornismya Fanny, Less.
Trochilus Labrador, Bourc.
Calothorax Fanny, Gray & Mitch.
Thaumastura fanny, Bonap.
Lucifer labrador, Bonap.
——— *Labrador*, Reichenb.
Myrtis Fanny, Reich.
Habitat. Ecuador and Peru.

Mr. Fraser found it at Cuenca in November 1857, when it was "common about the gardens and lane hedgerows. It makes much more humming with its wings than the long-tailed green *Lesbia*."— *Proc. Zool. Soc.* part xxvi. p. 459.

171. MYRTIS YARRELLI.
Calothorax Yarrelli . . . . Vol. III. Pl. CLII.
Trochilus Yarrelli, Bourc.
Habitat. Western Peru, particularly the neighbourhood of Arica.

How very singular and diversified in form are the Humming-Birds of Peru! So varied indeed are they that almost every species demands a generic appellation; the *T. Coræ* with its beautiful throat and lengthened tail is the type of the

Genus THAUMASTURA, *Bonap.*

Of this peculiar form only one species is known, of which the male alone is decorated with fine colours, and bears the singularly constructed tail.

172. THAUMASTURA CORÆ. . . . . . . Vol. III. Pl. CLIII.
Ornismya Cora, Less.
Calothorax cora, Gray & Mitch.
Trochilus Cora, Jard.
Thaumastura cora, Bonap.
Ornismya cora, Dev. Rev. et Mag. de Zool. 1852, p. 217.
Habitat. Peru.

M. Deville states that this bird is found during the months of February, March, April, and May only, in the humid districts bordering the River Rimæ at Lima. It is seen in small troops composed of six or eight couples, which are constantly pursuing one another, and uttering a slight cry. It is very airy in its flight, and rarely permits any other Humming-Bird to remain in its neighbourhood, but wages a continual and terrible war with them.

The largest species of the luminous lilac-throated Peruvian

Humming-Birds, the *Vespera*, constitutes the

Genus RHODOPIS, *Reichenb.*,

which, as the form differs from all the others, I have no other alternative, if I act with consistency, than to adopt. It has a deeply forked tail, the feathers of which are narrow and rigid, not soft and yielding as in the *T. Coræ*. I have never seen a second species of this particular form. The female, like that sex in many other allied genera, is destitute of brilliant colouring.

173. RHODOPIS VESPERA . . . . . Vol. III. Pl. CLIV.

Ornismya vesper, Less.
Trochilus vesper, Jard.
Calothorax vesper, Gray & Mitch.
Thaumastura vesper, Bonap.
Lucifer vesper, Bonap.
Rhodopis vespera, Reichenb.
Calliphlox vespera, Reichenb.

Habitat. Peru.

Not less beautiful in the colouring of their gorgets are the members of the

Genus DORICHA, *Reichenb.*

The *Eliza*, the Guatemalan bird known as *enicura*, and the less-known Bahama species *Evelynæ*, are all associated by me in this genus; and if the Plates on which they are respectively figured be referred to, it will be seen how beautiful are the throat-markings of the males.

174. DORICHA ELIZÆ.

Thaumastura Elizæ . . . . Vol. III. Pl. CLV.
Trochilus Eliza, Less.
Myrtis Eliza, Reichenb.
Lucifer eliza, Bonap.
Calothorax Eliza, Gray & Mitch.
Thaumastura Eliza, Montes de Oca in Proc. Acad. Nat. Sci. Philad. 1860, p. 552.

Habitat. Southern Mexico.

"This," says M. Montes de Oca, "is one of the rarest of the Mexican Humming-Birds. It is small, very beautiful, and flies with wonderful rapidity, moving its wings with such velocity that it is almost impossible to see them; and it might easily be mistaken for a large bee, from the buzzing sound produced by their incessant motion. In the vicinity of Jalapa it is called *Mirto de Colo de tierra*, or the spear-tailed Myrtle-sucker. It is very shy, and differs in its habits and manners from all other species. It is on the wing very early in the morning; and I have never seen any of the few specimens that have come under my observation between the hours of seven or eight o'clock A.M and five P.M., when they are again to be met with until dusk. When it has once been detected feeding at any particular spot, it is almost sure to be found there at the same hour for several days in succession. It feeds on the *Masapan* and *Tobaco* flowers, preferring, I think, the latter. It is also found and breeds at the Barrancas de Jico (or the Precipices of Jico), about twenty miles from Jalapa. The nest, which is very small, round, flat at the bottom, and neither so deep nor so thick at the base as those of most Humming-Birds, is covered on the outside with moss from stones, and lined with *tule*, or cattail silky floss."

175. DORICHA EVELYNÆ.

Thaumastura Evelynæ . . . . Vol. III. Pl. CLVI.

Trochilus Evelynæ, Bourc.
Calothorax Evelynæ, Gray & Mitch.
———— *Evelinæ*, Reichenb.
Callothorax evillina, Bonap.
Trochilus Bahamensis, Bryant.

Habitat. Bahama Islands.

176. DORICHA ENICURA.

Thaumastura enicura . . . . Vol. III. Pl. CLVII.

Trochilus enicurus, Vieill., Temm., Jard.
Ornismya heteropygia, Less.
Trochilus Swainsonii, Less.
Calothorax enicurus, Bonap.

Habitat. Guatemala.

"On no occasion," says Mr. Salvin, "were the males of this species observed about Dueñas during the months of February and March; indeed it was not until the month of May that both males and females were seen together, at which time the *Nopal* of the cochineal plantations being in full flower, great numbers of Humming-Birds, especially of this species, were in the habit of feeding from the blossoms of that cactus. The females during the winter months are common enough, and frequent the same places, and feed principally on the same trees as the *Cyanomyia cyanocephala*."—*Ibis*, vol. i. p. 129.

"Occasionally, when flying, the elongated tail-feathers are stretched to a considerable angle."—*Ibis*, vol. ii. p. 40.

Speaking of three nests of this species Mr. Salvin says:—"One of these was in a coffee-tree, and had two eggs. The other was most curiously placed in the cup-shaped top of a fruit of the Nopal (*Cactus cochinellifer*), the fastenings being dexterously wound round the clustering prickles, and thus retaining the whole structure most firmly in its place. This nest was remarkably shallow; so much so that, if it had not contained its two eggs, I should have pronounced it far from complete. It may be that, being based on a firm foundation (one not nearly so liable to oscillation by the wind) the bird had found that a greater depth was not necessary to keep the eggs from falling out. Had she placed her nest on a slender twig, as seems to be usual, the case might have been different. The third nest had young. It was placed in the upper shoots of a Dahlia at the further end of the court-yard. The hen seemed to have the entire duty of rearing the young; for I never once saw the male near the place; in fact, I never saw a male inside the court-yard. When sitting she would sometimes allow me to go close to her, and even hold the branch still when it was swaying to and fro by the wind, without evincing the slightest alarm. But it was only when a hot sun was shining that she would allow me to do this; when it was dull or raining four or five yards was the nearest I could approach. Frequently when I had disturbed her, I would sit down close at hand and wait for her return, and I always noticed that, after flying past once or twice overhead, she would bring a small piece of lichen, which, after she had settled herself comfortably in her nest, she would attach to the outside. All this was done with such a confident and fearless air, that she seemed to intimate, 'I left my nest purely to seek for this piece of lichen, and not because I was afraid of you.' When sitting upon her nest, the whole cavity was quite filled by her puffed-out feathers, the wings, with the exception of their tips, being *entirely concealed* by the feathers of the back. When the young were first hatched, they looked little, black, shapeless things, with long necks and hardly any beak. They soon, however, grew, and entirely filled the nest. I never saw the old bird sitting after the young had emerged from the eggs: she seemed to leave them alike in sun and rain. When feeding them she would stand upon the edge of the nest with her body very upright. The first of these young ones flew on October 15. It was standing on the side of the nest as I happened to approach, when it immediately flew off, but fell among the flowers below. I placed it in the nest, but a moment after it was off again, nothing daunted by its first failure—this second time with better success, for it flew over a wall close by and settled on a tree on the other side. In the evening I saw the old one feeding it, and went up to the tree; but it started off with increased vigour to an orange-tree, and tried at first to rest on one of the fruit, but failing, found a more appropriate perch on the edge of a leaf. I never saw it afterwards. The other young one flew two days later.

"The seeds of the willow and bulrush are favourite materials for the interior structure of the nest, while lichen is freely used outside." —*Ibis*, vol. ii. p. 264.

Genus TRYPHÆNA, *Gould.*

(Τρύφαινα, nom. prop.)

Generic characters.

Male.—*Bill* as long as the head, and straight; *wings* very small; *primaries* narrow; *tail* deeply forked, the outer feather narrow, tapering at the tip and incurved; *feet* small, claws short and hooked; gorget richly coloured but not luminous; tail ornamented.

Female.—Unadorned; *tail* extremely short.

The single species of this genus stands quite alone in the great family of Humming-Birds. The peculiar and beautiful markings of its tail are most remarkable; the colouring of the throat-mark is equally distinct. It must be remembered that these features are confined to the male, the female being very plainly attired, and having a very diminutive tail. Guatemala may well be proud of this singular bird, rich as her fauna really is.

177. TRYPHÆNA DUPONTI . . . . . . Vol. III. Pl. CLVIII.

Ornismya Duponti, Less., Jard.
——— *Zémès*, Less.
——— *cœlestis*, Less.
Mellisuga Duponti, Gray & Mitch.
**Trochilus Duponti*, Jard. Nat. Lib. Humming-Birds, vol. i. p. 131. pl. 26.
**Cynanthus Duponti*, Ib. vol. ii. p. 145.
**Trochilus lepidus*, Licht. in Mus. Berol.
*Tilmatura lepida, Reichenb. Aufz. der Col. p. 8; Ib. Troch. Enum. p. 5. pl. 711. figs. 4610, 4611.
**Thaumastura Duponti*, Bonap. in Rev. et Mag. de Zool. 1854, p. 257.
*——— ——— *Duponti*, Cab. et Hein. Mus. Hein. Theil iii. p. 55, note.

Habitat. Guatemala.

" San Gerónimo, December 10. Don Vicente Constancia assures me that this species is found near the city of Guatemala; otherwise this is the only locality I have been able to discover, as yet, where it occurs.

" Following the course of the river of San Gerónimo up its bed to about half a league from the village, you come upon a small patch of forest with here and there open spots covered with *Salviæ*. Here it was that this bird was shot by a boy, who told me there were plenty; however on visiting the place soon after, I was not successful in obtaining more specimens, nor was I fortunate enough to see one."—*Salvin* in *Ibis*, vol. ii. p. 266.

Genus CALLIPHLOX, *Boié*.

Perhaps the very commonest of the frill-necked Humming-Birds is the *C. amethystina*. It is more widely spread than many other species, since it inhabits all the countries from Brazil to Venezuela.

In this genus I have also placed the *C. Mitchelli*; but I have some doubt as to the propriety of so doing. The throats of the two birds, although beautifully coloured, are not luminous.

178. CALLIPHLOX AMETHYSTINA . . . Vol. III. Pl. CLIX.

Trochilus amethystinus, Gmel., Lath., Vieill., P. Max. zu Wied, Shaw, Jard.
Ornismya amethystina, Less.
——— *orthura*, Less. ?

Mellisuga amethystina, Steph., Gray & Mitch.
Trochilus campestris, Pr. Max. zu Wied.
Tryphæna amethystinus, Bonap.
——— *amethystina*, Bonap.
Calliphlox amethystina, Reichenb.
Amethystine Humming-Bird, Lath., Shaw.
Tryphæna amethystina, Gould.
* *Trochilus brevicaudus*, Spix, Av. Bras. tom. i. p. 79. tab.
* ——— *orthurus*, Jard. Nat. Lib. Humming-Birds, vol. ii. p. 60. pl. 8.?
* *Cynanthus amethystinus*, Ib. vol. ii. p. 143.
* ———? *orthura*, Ib. vol. ii. p. 143.?

Habitat. Brazil, Demerara, Cayenne, and Trinidad.

Whether the *Ornismya orthura* of Lesson be a species or an old female of *C. amethystina* requires further investigation. Wherever the *C. amethystina* is found in Brazil, Trinidad, or Demerara, the *O. orthura* is found in its company—a fact which militates against its being a distinct species.

179. CALLIPHLOX AMETHYSTOÏDES, *Less.*

Ornismya Amethystoïdes, Less.
Mellisuga Amethystoïdes, Gray & Mitch.
* *Trochilus amethystoides*, Jard. Nat. Lib. Humming-Birds, vol. ii. p. 62.
* *Cynanthus amethystoides*, Ib. vol. ii. p. 143.
* *Calliphlox amethystoides*, Bonap. Consp. Gen. Av. p. 84.

Habitat. Minas Geraes in Brazil.

In my account of *C. Amethystina* I have regarded this species as identical with that bird; but M. Bourcier is still of opinion that it is distinct; and as I find that it is of smaller size, and the tint of the gorget is somewhat different, I defer to his opinion. It will not, however, be necessary to give a separate figure of it. Specimens from Minas Geraes are certainly more diminutive than from elsewhere.

180. CALLIPHLOX ? MITCHELLI . . . . . Vol. III. pl. CLX.

Trochilus Mitchelli, Bourc.
Mellisuga Mitchelli, Gray & Mitch.
Calothorax Mitchelli, Reichenb.
Tryphæna mitchelli, Bonap.

Habitat. Southern parts of New Grenada and Ecuador.

I have now gone through the species of Humming-Birds distinguished for their diminutive size, the delicacy of their structure, and for their luminous gorgets. It is true that many other groups have their throats similarly adorned, such as the members of the genera *Oreotrochilus, Heliangelus*, &c.; but these birds are all of large size and of very different form, and as we proceed I think it will be seen that they are better placed hereafter. I proceed next, then, with the racquet-tailed species—the *Spathuræ*, &c. I admit that there is no direct alliance between these and the former, but it will be recollected that I have stated that the Humming-Birds cannot be arranged in anything like a series of affinities. Here, then, we commence with a very different group; remarkable for the peculiar character of the tail in most of its members. Among these I place in the foremost rank the extraordinary bird bearing the specific name *mirabilis*. As any description, however accurate, must fail to give a correct idea of this singular species, I must refer my readers to the Plate, upon which it is correctly depicted.

Genus LODDIGESIA, *Gould*.
(*Loddiges*, nom. prop.)

Male.—Bill straight and longer than the head; *wings* diminutive; *primaries* rounded at the tip; outer tail-feather on each side very much prolonged, and terminating in a large spatule.
Female.—Unknown.

181. LODDIGESIA MIRABILIS, *Gould* . . Vol. III. Pl. CLXI.

Trochilus mirabilis, Lodd. MS., Bourc.
Loddigesiornis mirabilis, Bonap.

Habitat. Chachapoyas in Peru.

The racquet-tailed birds I have figured under the generic name of *Spathura* are spread over the temperate regions of the great Andean range of mountains from the northern parts of New Granada to Bolivia. Much confusion prevails with respect to the generic appellation of these birds. The case stands thus: in 1846 I proposed the term *Ocreatus* for the *rufocaligatus*, and in 1850 substituted that of *Spathura*; while in 1849 Dr. Reichenbach employed that of *Steganurus*, which he changed in 1853 to *Steganura*. In the body of this work all the species are arranged under my own generic name, which I hope may be allowed to stand.

Genus SPATHURA, *Gould*.
(Σπάθη, spatha, et οὐρά, cauda.)

Generic characters.
Male.—Bill straight and rather longer than the head; *wings* moderately long and somewhat rounded; *tail* deeply forked; the outer tail-feather on each side terminating in a spatule; *feet* small; *tarsi* thickly clothed; *hind toe* and *nail* shorter than the middle toe and nail; *throat* luminous.
Female.—Unadorned and destitute of spatules.

182. SPATHURA UNDERWOODI . . . . Vol. III. Pl. CLXII.

Ornismya Underwoodi, Less.
Trochilus Underwoodi, Jard.
Trochilus ventilabrum, Lath.
Mellisuga Underwoodi, Gray & Mitch.

*_Ornismya Kieneri_, Less. Les Troch. p. 165, pl. 65, female.
*_Cynanthus Underwoodi_, Jard. Nat. Lib. Humming Birds, vol. ii. p. 144.
*———— _Kienerii_, Ib. p. 146.
*_Steganura spatuligera_, Reich. Aufs. der Col. pp. 8 & 24; Id. Troch. Enum. p. 5. pl. 708. figs. 4598–600.
*———— _Underwoodi_, Cab. et Hein. Mus. Hein. Theil iii. p. 66.
*_Steganura remigera_, Reichenb. Aufs. der Col. pp. 8, 24; Id. Troch. Enum. p. 5, pl. 708, figs. 4601–2.
*_Steganurus remigera_, Cab. et Hein. Mus. Hein. Theil iii. p. 67.

Habitat. The neighbourhood of Bogota, on the Andes, and the hilly portion of eastern Venezuela.

In his 'Trochilinarum Enumeratio' Dr. Reichenbach has figured a white-booted Racquet-tail under the name of _Steganura remigera_, which, after a careful examination of the type specimen, I have no doubt is identical with this species, I have therefore placed that name among its synonyms.

183. SPATHURA MELANANTHERA, _Jard._ . Vol. III. Pl. CLXIII.

Trochilus (Spathura) melananthera, Jard.
Steganura melananthera, Reichenb.
Discura melananthera, Bonap.

Habitat. Ecuador.

Mr. Fraser, who procured specimens of _S. melananthera_ at Pallatanga and Nanegal in Ecuador, states that its feet are "white."

184. SPATHURA PERUANA, _Gould_ . . . Vol. III. Pl. CLXIV.

Habitat. Moyobamba in Peru.

185. SPATHURA RUFOCALIGATA, _Gould_ . Vol. III. Pl. CLXV.

Trochilus (Ocreatus) rufocaligatus, Gould.
———— _Addæ_, Bourc.
Mellisuga rufocaligata, Gray & Mitch.

Habitat. La Paz in Bolivia.

The _Trochilus Addæ_ of M. Bourcier is considered to be identical either with the _S. Peruana_ or the present bird; in all probability it was applied to the latter; and if this should prove to be the case, the term _Addæ_, having been proposed prior to that of _rufocaligata_, should be adopted for this species.

186. SPATHURA CISSIURA, _Gould_ . . . Vol. III. Pl. CLXVI.

Habitat. Peru.

Perhaps the next in affinity, although not directly allied, are the members of the genus _Lesbia_, which are equally confined to the Andes, and fly at the same elevation as the _Spathuræ_. Like those birds, they are distributed along that great chain of mountains throughout many degrees of latitude on each side of the equator. They may be regarded as among the most elegant of the Trochilidæ. Their long and deeply cleft tails would seem to indicate that they possess very great powers of aërial progression,—a remark which equally applies to the members of the genera *Cynanthus* and *Cometes*.

Genus LESBIA, *Less.*

187. LESBIA GOULDI . . . . . . Vol. III. Pl. CLXVII.

Trochilus Gouldii, Lodd.
Ornismya sylphia, Less.
Mellisuga Gouldii, Gray & Mitch.
Cynanthus gouldi, Bonap.
Lesbia Gouldii, Reichenb.
Agaclyta Gouldi, Cab. et Hein. Mus. Hein. Theil iii. p. 70.

Habitat. The high lands of New Granada, particularly the neighbourhood of Bogota.

188. LESBIA GRACILIS, *Gould* . . . . Vol. III. Pl. CLXVIII.

Trochilus (Lesbia) gracilis, Gould.
Mellisuga gracilis, Gray & Mitch.
Cynanthus gracilis, Bonap.
Lesbia gracilis, Reichenb.

Habitat. Ecuador.

189. LESBIA NUNA . . . . Vol. III. Pl. CLXIX.

Ornismya Nuna, Less.?

Habitat. Peru.

Refer to my remarks respecting this bird in the letter-press accompanying the Plate.

190. LESBIA AMARYLLIS . . . . . . Vol. III. Pl. CLXX.

Trochilus Amaryllis, Bourc. et Muls., Gray & Mitch.
Cynanthus amaryllis, Bonap.
"*Trochilus Victoriæ*, Bourc. Rev. Zool. 1846, p. 315, pl. 4.
"*Mellisuga Victoriæ*, Gray & Mitch. Gen. of Birds, vol. i. p. 103, *Mellisuga*, sp. 54.
"*Cynanthus Victoriæ*, Bonap. Consp. Gen. Av. tom. i. p. 81. *Cynanthus*, sp. 6.
"*Lesbia Victoriæ*, Reich. Aufz. der Col. p. 8; Id. Troch. Enum. p. 5, pl. 715. figs. 4622–23.
"*Psalidoprymna Victoriae*, Cab. et Hein. Mus. Hein. Theil iii. p. 52.

Habitat. New Granada, Ecuador, and Peru.

Dark or nearly black varieties not unfrequently occur among the Trochilidæ; and I think that the bird to which M. Bourcier has given the name of *Victoriæ* is merely such a variety of the *L. Amaryllis*.

191. LESBIA EUCHARIS . . . . Vol. III. Pl. CLXXI.
Trochilus eucharis, Bonrc.
Lesbia eucharis, Reichenb.
Cynanthus eucharis, Bonap.
*Lesbia bifurcata, Reich. Troch. Enum. pl. 716. figs. 4624–25.
Habitat. New Granada.

Considerable, and I fear inextricable, confusion exists with regard to the genera *Lesbia* and *Cynanthus*, which would appear to be due to the various authors who have used those terms taking their characters from defective descriptions or imperfect drawings, instead of actual specimens. This confusion I have endeavoured to rectify by applying the terms to the birds which I believe their respective proposers actually intended, and I do hope that, for the sake of science, they will be allowed to to stand for the future.

Leaving the genus *Lesbia*, then, we proceed to that of *Cynanthus*, and here we arrive at some of the most remarkable and the most beautiful of the Trochilidæ. Strictly confined to the great Andean mountains and the spurs which jut out as far as eastern Venezuela, these blue-tailed birds enjoy a range of habitat extending from the lands washed by the Caribbean Sea to Peru.

Those inhabiting the neighbourhood of Bogota appear to be divided into two or three local varieties or races; for they are not, in my opinion, sufficiently different to warrant us in regarding them as species. On the other hand, the Ecuadorian bird possesses characters which induced me to consider it distinct.

The variation observable among the Bogotan birds is principally in the colouring of the tail—some having the whole of the feathers blue, while others have the eight central ones tipped with beautiful green.

Genus CYNANTHUS, *Swains.*

192. CYNANTHUS CYANURUS . . . . Vol. III. Pl. CLXXII.
Trochilus cyanurus, Steph.
Orniomya Kingii, Less.
**Lesbia forficatus*, Reich. Aufz. der Col. p. 8; Id. Troch. Enum. p. 5, pl. 718. figs. 4628–29.
Lesbia Gorgo, Reich. Aufz. der Col. pp. 8, 24; Id. Troch. Enum. p. 5; Cab. et Hein. Mus. Hein. Theil iii. p. 71.
Habitat. New Granada.

A somewhat smaller and more delicate bird than the *Cynanthus cyanurus* occurs in Venezuela, having the whole of the body green, with the exception of a patch of blue on the throat; and the crown brilliant metallic green, without the superciliary stripe of black seen in that species: I refrain, however, for the present from characterizing it as distinct.

193. CYNANTHUS CŒLESTIS, *Gould.*
Habitat. Ecuador.

This new Humming-Bird is considerably larger than the *C. cyanurus*; it also presents a marked difference in the colouring of the under surface, which is uniform coppery brown, instead of green; in other respects the colouring is very similar to the specimens from Bogota, with green and blue tails. In no instance have I seen a specimen from Ecuador with an entirely blue tail, whereas they frequently occur among those sent from Bogota.

194. CYNANTHUS MOCOA.

Cynanthus smaragdicaudus, *Gould* . . Vol. III. Pl. CLXXIII.
Trochilus Mocoa, Delatt. et Bourc.
———— (*Lesbia*) *smaragdinus*, Gould.
Mellisuga smaragdinis, Gray & Mitch.
Cynanthus mocoa, Bonap.
Habitat. Peru and Bolivia.

Specimens of this species, like those of the *C. cyanurus*, are found to differ considerably; but as it is a bird of comparative rarity, we have seen too few examples to come to any positive conclusion as to whether these are referable to one or two species. The *Mocoa* may be regarded as the southern representative of the *C. cyanurus*. It frequents the forests of Bolivia and Peru, particularly those clothing the eastern slopes of the great Andean range.

As the *Lesbiæ* naturally led us on to the *Cynanthi*, so do the latter in their broad tail-feathers offer an alliance to the *Cometæ*; and, however much I have extolled the beauty of any of the preceding genera, it is scarcely possible to select terms sufficiently expressive to convey an idea of the loveliness of the birds comprised in this latter genus. The two birds generally known under the names of *Sappho* and *Phaon* are *par excellence* the most gorgeous birds in existence so far as regards the colouring of their tails; and well do these living meteors deserve the more general name of *Comets*.

Genus COMETES, *Gould*.
(Κομήτης, cometa.)

Generic characters.
Male.—*Bill* longer than the head, straight or slightly arched; *wings* moderate; *tail* long and deeply forked, the feathers broad and luminous; *tarsi* naked; *feet* small; *hind toe* and *nail* nearly as long as the middle toe and nail; *throat* luminous.
Female.—Smaller in size and nearly destitute of fine colouring.

195. COMETES SPARGANURUS . . . . Vol. III. Pl. CLXXIV.

Trochilus sparganurus, Shaw, Steph., Jard.
———— *chrysurus*, Cuv.
———— *radiosus*, Temm.
Ornismya Sappho, Less.
Cometes Sappho, Gould.
———— *sparganurus*, Bonap.
Mellisuga sparganura, Gray & Mitch.

Orthorhynchus chrysurus, D'Orb. et Lafres.
Trochilus chrysochloris, Vieill.
Cynanthus sparganurus, Jard. Nat. Lib. Humming Birds, vol. ii.
p. 146.
Trochilus (Cynanthus) chrysurus, Tsch. Consp. p. 36, sp. 200;
Id. Faun. Per. p. 244.
Orthorhynchus chrysurus, D'Orb. et Lafres. Syn. p. 26.
Sappho sparganura, Reich. Aufz. der Col. p. 9; Id. Troch. Enum.
p. 5, pl. 724. figs. 1651-52.
Lesbia sparganura, Bonap. Rev. et Mag. de Zool. 1854, p. 252.
Sparganura Sappho, Cab. et Hein. Mus. Hein. Theil iii. p. 52.
Habitat. Bolivia.

196. COMETES PHAON, *Gould* . . . . Vol. III. Pl. CLXXV.

Mellisuga phaon, Gray & Mitch. Gen. of Birds, vol. i. p. 113,
Mellisuga, sp. 47.
Cometes phaon, Bonap.
Sappho phaon, Reich. Aufz. der Col. p. 9; Id. Troch. Enum. p. 5,
pl. 725. figs. 1653-54.
Lesbia phaon, Bonap. Rev. et Mag. de Zool. 1854, p. 252.
Sparganurus Phaon, Cab. et Hein. Mus. Hein.Theil iii. p. 52, note.
Habitat. Peru.

197. COMETES? GLYCERIA . . . . Vol. III. Pl. CLXXVI.

Cometes Mossai, Gould.
Lesbia glyceria, Bonap.
Sparganura Mossai, Cab. et Hein. Mus. Hein. Theil iii. p. 52, note.
Habitat. Popayan in New Granada.

This is perhaps the most extraordinary bird I have had the good fortune to describe. I have placed it in the genus *Cometes* with a reservation; for it comprises characters which are found both in *Lesbia* and *Cometes*: in form it most nearly approaches the latter, while in its markings and in the colouring of its throat it resembles the former. At present only a single example has been obtained, and this, I am inclined to think, is not quite adult; it may possibly be only a young male of a splendidly coloured but unknown species; and if so, a fine bird remains in store to reward the researches of some future explorer.

198. COMETES? CAROLI . . . . . Vol. III. Pl. CLXXVII.

Trochilus Caroli, Bourc.
Hylocharis Caroli, Gray & Mitch., Bonap.
Calliphlox Caroli, Reichenb.
Avocettinus carolus, Bonap.
Habitat. Peru.

Of this remarkable bird about four specimens have been in our collections for many years; but whether they are males or females is unknown; for in fact nothing has been recorded respecting these puzzling birds. If the description accompanying my plate of the species be referred to, it will be seen that it has been bandied about from one genus to another, different authors having assigned it to *Trochilus, Hylocharis, Calliphlox,* and *Avocettinus*! Some day, when the little-known country of Peru has been more fully investigated, we shall doubtless acquire a better knowledge of it, and be able to decide to which genus it really pertains: for the present let it remain in the one in which I have placed it.

Genus PTEROPHANES, *Gould*.

(Πτερὸν, ala, et φαίνω, ostendo.)

Generic characters.

Male.—*Bill* cylindrical, longer than the head, and slightly upcurved; *wings* very large and sickle-shaped; *tail* broad and large; *tarsi* clothed; *feet* small; *hind toe* shorter than the middle one; *claws* long, slightly curved, and sharp at the point.

Female.—Unadorned.

199. PTEROPHANES TEMMINCKI . . Vol. III. Pl. CLXXVIII.

Ornismya Temmincki, Boiss.
Trochilus cyanopterus, Lodd. MS.
Mellisuga Temmincki, Gray & Mitch.
**Pterophanes Temmincki,* Reichenb. Aufz. der Col. p. 14; Id. Troch. Enum. p. 11; Cab. et Hein. Mus. Hein. Theil iii. p. 80.

Habitat. New Granada and Ecuador.

The *Pterophanes Temmincki* must rank with the *Patagona gigas* among the very largest of the Humming-Birds; the two species are nearly equal in size, but in their structure and the colouring of their plumage they are very different. The native country of the *P. Temmincki* is the temperate portion of the Andes, over which it ranges for a considerable distance from Bogota, the probable centre of its area. I have also seen specimens from Ecuador, where it appears to be scarce. This fine bird is rendered a very striking species by the beautiful blue colouring of its wings.

Genus AGLÆACTIS, *Gould*.

('Αγλαΐα, splendor, et ἀκτίς, radius solaris.)

Generic characters.

Male.—*Bill* rather short, depressed at the base, and straight; *wings* long and powerful; *primaries,* particularly the outer one, sickle-shaped; *tail* moderately large, and slightly forked when closed; *tarsi* partially clothed; *feet* strong and powerful; *hind toe* and *nail* longer than the middle toe and nail; *breast* ornamented with a tuft of lengthened plumes; *back* luminous when viewed from behind.

The birds for which I instituted the above genus have always greatly interested me. They are of large size, have very ample wings, and are distinguished from all other Humming-Birds by their luminous backs, of which the rich and glittering hues are only perceptible when viewed from behind, or reversely to the direction of the feathers; contrary to the law which regulates the disposition of the colouring in all the other genera except in *Cæligena*, where it is slightly apparent. All the species are natives of the Andes, over which they roam from the northern part of New Granada to Bolivia. The latter country is the cradle of the *Aglæactis Castelnaudi* and the richly coloured *A. Pamela*. These extraordinary birds, to which I have given the trivial name of Sunbeams, are among the most wonderful of the Trochilidæ.

200. AGLÆACTIS CUPRIPENNIS . . . Vol. III. Pl. CLXXIX.

Trochilus cupripennis, Bourc. et Muls.
Mellisuga cupripennis, Gray & Mitch.
Aglæactis cupripennis, Bonap.
―――― *cupreipennis*, Bonap.
Helianthea cupripennis, Reichenb.
Aglaiactis cupripennis, Cab. et Hein. Mus. Hein. Theil iii. p. 69.

Habitat. New Granada.

Professor Jameson and Mr. Fraser state that "The females of this species have the glittering back, but not so brilliant as in the males."
—*Ibis*, vol. i. p. 400.

201. AGLÆACTIS ÆQUATORIALIS.

Aglaiactis æquatorialis, Cab. et Hein. Mus. Hein. Theil iii. p. 70, note.

Habitat. Ecuador.

This bird is considered distinct from *H. cupripennis* by Dr. Cabanis; but the only difference I can perceive between this and Bogota specimens is in its larger size and longer wing; but if this be admitted as a species, I must further increase the list by calling another *parvula*, of which I have two specimens shot by M. Warszewicz in Peru or Bolivia, for the precise locality is unknown to me.

202. AGLÆACTIS PARVULA, *Gould.*

Habitat. Peru, or Bolivia.

This bird has a much shorter bill, is of a deeper red on the under surface, more red in the tail, and altogether of much smaller size.
Total length 4¼ inches; bill $\frac{11}{12}$; wing 3; tail 1¼.

203. AGLÆACTIS CAUMATONOTA, *Gould.*

Aglæactis caumatonotus, Gould in Proc. Zool. Soc. part xvi. 1848, p. 12.

Habitat. Peru, or Bolivia.

Described by me as above from a single specimen said to have been procured in Peru, which differs from the preceding in being of smaller size and in the darker hue of the luminous portion of the back.

204. AGLÆACTIS CASTELNAUDI . . . Vol. III. Pl. CLXXX.

Trochilus Castelnaudii, Bourc. et Muls.
——— *Castelnaui*, Gray & Mitch.
Aglaeactis castelnaudi, Bonap.
Aglæactis castelneaui, Bonap.
*Aglaeactis Castelnaudi, Reichenb. Aufs. der Col. p. 9.
*Helianthea Castelnaudi, Reichenb. Troch. Enum. p. 6, pl. 739. figs. 4694–95.
*AglalactisCastelnaui, Cab. et Hein. Mus. Hein. Theil iii. p. 69, note.
*Ornismya Castelnandii, Dev. Rev. et Mag. de Zool. 1852, p. 216.

Habitat. The mountains near Cusco in *Peru*.

M. Deville says "this bird, which is very rare, confines itself to the blossoms of a species of mimosa, the odours of which attract the small insects which form its food. Its cry is very piercing; its flight very rapid and noisy. This species, which is quite new, was killed by myself in the valley of Echaraté, near Cusco."

205. AGLÆACTIS PAMELA . . . . . Vol. III. Pl. CLXXXI.

Orthorhynchus Pamela, D'Orb. et Lafres.
Hyloeharis Pamela, Gray & Mitch.
Aglæetis pamela, Bonap.
Helianthea Pamela, Reichenb.
*Aglaiactis Pamelae, Cab. et Hein. Mus. Hein. Theil iii. p. 69.

Habitat. Bolivia.

Distinct from every other genus are the two species of *Oxypogon*. These bearded birds stand quite alone among the Trochilidæ; and although not remarkable for brilliancy of colour, their fantastic markings, towering crests, and lengthened beards render them very conspicuous objects. I shall not be surprised if other species of this form be discovered when the higher peaks of the great Andean range of mountains have been more closely examined.

For a long time the *Oxypogon Guerini* was the only species known; but in the year 1842 the intrepid traveller Mons. J. Linden ascended the high mountains of the Sierra Nevada de Merida, and was rewarded by the discovery of the second species, which bears his name.

Genus OXYPOGON, *Gould.*

('Ὀξύς, acutus, et πώγων, barba.)

Generic characters.

Male.—Bill shorter than the head, feeble, and straight; *face* both above and below ornamented with lengthened plumes, the former erect, the latter pendent; *wings* rather long; *tail* large and forked when closed; *tarsi* bare; *feet* large and strong; *hind toe* and *nail* longer than the middle toe and nail.

*Female.—*Smaller, and destitute of the ornamental face-plumes.

206. OXYPOGON GUERINI . . . . Vol. III. Pl. CLXXXII.
Ornismia Guerinii, Boiss.
Trochilus parvirostris, Fras.
Ornismya Guerinii, Lodd.
Mellisuga Guerinii, Gray & Mitch.
Oxypogon Guerini, Reich. Aufz. der Col. p. 12; Id. Troch. Enum. p. 10; Cab. et Hein. Mus. Hein. Theil iii. p. 67.

Habitat. New Granada; plentiful around Bogota.

207. OXYPOGON LINDENI . . . . Vol. III. Pl. CLXXXIII.
Ornysmia Lindenii, Parz.
Mellisuga Lindenii, Gray & Mitch.
Oxypogon Lindeni, Heichenh. Aufz. der Col. p. 12; Id. Troch. Enum. p. 10; Cab. et Hein. Mus. Hein. Theil iii. p. 67, note.

Habitat. The Sierra Nevada de Merida in New Granada.

"This bird," says M. Linden, "inhabits the regions immediately beneath the line of perpetual congelation, and never at a less elevation than 9000 feet." It might be thought that such bleak and inclement situations were ill-adapted for so delicate a structure as that of the Humming-Bird; but there, and there only, does it dwell, while the equally lofty Paramos of Bogota are the native locality of the allied species *O. Guerini*. The minute insects which frequent the alpine flora of these districts afford abundance of food to these birds; and beautifully constructed are their little bills for searching among the flowers in which they are found.

Near the members of the genus *Oxypogon* are the various species of *Ramphomicron*, another bearded group, but differing in the total absence of any lengthened plumes on the crown, and in the structure and colour of the pendent chin feathers. It will only be necessary to glance at the plates on which these species are depicted to perceive that, though they bear a general resemblance to the *Oxypogons*, they are generically distinct from them. Their short and feeble bills indicate that they feed on a similar kind of insect food; and we know that such flowers as those of *Sida* and other plants with open corollas are frequently visited for the insects which abound therein.

It is said that the members of this genus fly with great rapidity, and that like flashes of light they are constantly dashing about the hillsides from one flower to another. It must be extremely interesting to watch the aërial movements of these comparatively large birds among the lofty regions they frequent, and where the air is so pure and rarified. In all the hilly countries, from the Caribbean Sea southward to Bolivia, are the members of this genus to be obtained; in the neighborhood of Bogota one of them is very common; this bird (the *R. heteropogon*) extends its range from thence to about the latitude of Popayan, while the little *microrhyncha* is equally abundant in New Granada and Ecuador. At Quito, or around those towering mountains immediately under the equator, we find the *Stanleyi* and *Herrani*; while Bolivia gives us the *Vulcani* and the *ruficeps*.

Genus RAMPHOMICRON, *Bonap.*

208. RAMPHOMICRON HETEROPOGON . Vol. III. Pl. CLXXXIV.

Ornismya heteropogon, Boiss,
Trochilus coruscus, Fras.
Mellisuga heteropogon, Gray & Mitch.
Ramphomicron heteropogon, Bonap.
Lampropogon heteropogon, Bonap. Rev. & Mag. de Zool. 1854, p. 252.
Chalcostigma heteropogon, Reichenb. Aufz. der Col. p. 12.
Ramphomicron heteropogon, Reichenb. Troch. Enum. p. 10 ; Cab. et Hein. Mus. Hein. Theil iii. p. 67.

Habitat. New Granada.

I possess two very marked varieties or races of this bird, one being much smaller than the other: the large race, I believe, is from Pamplona, and the smaller from the neighbourhood of Bogota.

209. RAMPHOMICRON STANLEYI . . Vol. III. Pl. CLXXXV.

Trochilus Stanleyi, Bourc. et Muls.
Chalcostigma Stanleyi, Reichenb. Aufz. der Col. p. 12 ; Id. Troch. Enum. p. 10.

Habitat. Ecuador.

210. RAMPHOMICRON VULCANI, *Gould* Vol. III. Pl. CLXXXVI.

Chalcostigma Vulcani, Reichenb. Aufz. der Col. pl. 12.
Ramphomicron Vulcani, Reichenb. Troch. Enum. p. 10.

Habitat. Bolivia.

211. RAMPHOMICRON HERRANI . . Vol. III. Pl. CLXXXVII.

Trochilus Herrani, De Latt. & Bourc.
Calothorax herrani, Bonap.
Chalcostigma Herrani, Reichenb. Aufz. der Col. p. 12; Id. Troch. Enum. p. 10.
Lampropogon herrani, Bonap. Rev. et Mag. de Zool. 1854, p. 253.

Habitat. Ecuador.

212. RAMPHOMICRON RUFICEPS, *Gould.* Vol. III. Pl. CLXXXVIII.

Trochilus (——?) *ruficeps*, Gould.
Mellisuga ruficeps, Gray & Mitch.
Ramphomicron ruficeps, Bonap.
Ramphomicron ruficeps, Reichenb. Aufz. der Col. p. 12; Id. Troch. Enum. p. 10.
Lampropogon ruficeps, Bonap. Rev. et Mag. de Zool. 1854, p. 253.

Habitat. Bolivia.

213. RAMPHOMICRON MICRORHYNCHUS. Vol. III. Pl. CLXXXIX.

Ornismya microrhyncha, Boiss.
Trochilus brachyrhynchus, Fras.

Mellisuga microrhyncha, Gray & Mitch.
Ramphomicron microrhyncha, Bonap.
* *Rhamphomicron microrhynchum*, Reichenb. Auft. der Col. p. 12;
   Id. Troch. Enum. p. 10, pl. 718. figs. 4915-18.
* *Rhamphomicrus microrhynchus*, Cab. et Hein. Mus. Hein. Theil iii.
   p. 70.
* *Trochilus euanthes*, Licht. in Mus. of Berlin.
Habitat. The Andes from the equator to seven degrees north.

I must now ask those who take an interest in the various forms of this family of birds to turn to my plate of *Urosticte Benjamini*, and examine the little bird figured thereon with a beautiful gorget of green and purple. This species is rendered very singular by the two tufts of white feathers which spring from behind the eye, and still more so by the white tipping of the four central tail-feathers. Ornithologists will view this character with astonishment, and inwardly ask, Is this particular mark given for a special purpose in connexion with the economy of the bird, or for the mere purpose of ornament? That ornament and variety is the sole object, I have myself but little doubt. Of this recently acquired form, the single species to which I have assigned the generic name of *Urosticte* is all that is known. Like so many others that have preceded it, this is an Andean species, its native country being Ecuador.

Genus UROSTICTE, *Gould*.

(Οὐρά, cauda, et στικτός, notatus.)

Generic characters.

Male.—Bill much longer than the head, and straight; *head* round, the feathers not advancing on the bill; *wings* moderately long and rather pointed; *tail* slightly forked; *tarsi* clothed; *hind toe* shorter than the middle toe; *throat* luminous.

Female.—Unadorned.

214. UROSTICTE BENJAMINI . . . . . . Vol. III. Pl. CXC.

Trochilus Benjamini, Bourc.
* *Urosticta Benjamini*, Reichenb. Aufz. der Col. p. 13.
* *Basilinna Benjamini*, Reich. Troch. Enum. p. 11.
* *Urosticte benjaminus*, Bonap. Rev. et Mag. de Zool. 1854, p. 253.
Habitat. Ecuador.

It will have been observed that each of the foregoing groups is characterized by certain peculiarities, and that one feature is more prominent than the others in each of the different forms; in some the back, and back only, is lit up with luminous colours; in others, the throat is the only part thus adorned; in another (the *Pterophanes*) the wings alone are lustrous. The group which stands next on my list of genera and species have their share of ornament disposed on their broad and ample tails. In nearly every species this organ is illuminated with brilliantly shining colours—some blue, others purple, and others, again, bronze; in some these glittering hues appear on both the upper and under surface, while in others it is either confined to or is most brilliant on the latter. These colours, I am sure, the bird has the power of displaying to the greatest advantage, in order to render himself as attractive as may be when desirous of pleasing the female, perhaps, like the Peacock, for the purpose of his own vain glory. These varied beauties serve to increase our admiration of nature's works. At least such is the feeling they create in my own breast.

Genus METALLURA, *Gould.*

(Μέταλλον, metallum, et οὐρά, cauda.)

Generic characters.

Male.—*Bill* straight and of moderate length; *wings* moderate; *tail* rather large and rounded; *tarsi* bare; *feet* rather large; *hind toe* and *nail* as long or longer than the middle toe and nail; *throat* and *under surface* of the *tail* luminous, like shining metal.

Female.—Much less brilliant than the male, and in most of the species destitute of the luminous throat-mark.

All the members of this genus are tenants of the Andes, and by far the greater portion of them of Bolivia and Peru; one species, however (the *M. tyrianthina*), ranges over the whole of the temperate portions of New Granada. I now proceed to arrange the species according to their affinities, commencing with the largest and most gorgeously coloured.

215. METALLURA CUPREICAUDA, *Gould* . Vol. III. Pl. CXCI.

Trochilus (——?) *cupricauda*, Gould.
Mellisuga cupreo-cauda, Gray & Mitch.
Metallura cupreicaudus, Bonap.
—————— *cupreicauda*, Reichenb.
Aglæactis cupreicauda, Bonap.
Trochilus (Lampornis) opaca, "Licht." Tschudi, Consp. pp. 30, 210; Id. Faun. Peru. p. 248, 13.
Metallura opaca, Cab. et Hein. Mus. Hein. Theil iii. p. 69.

Habitat. Bolivia.

In the third part of his "Museum Heineanum," Dr. Cabanis has placed the specific name of *opaca* to this species as having the priority; if this should prove to be correct, my name of *cupreicauda* must sink into a synonym.

216. METALLURA ÆNEICAUDA, *Gould*. . Vol. III. Pl. CXCII.

Trochilus (——?) *æneocauda*, Gould.
Mellisuga æneocauda, Gray & Mitch.
Metallura æneicaudus, Bonap.
—————— *æneicauda*, Reichenb.
Aglæactis æneicauda, Bonap.

Urolampra aeneicauda, Cab. et Hein. Mus. Hein. Theil iii. p. 68.
Habitat. Bolivia.

217. METALLURA WILLIAMI . . . . Vol. III. Pl. CXCIII.
Trochilus Williami, Bourc. et De Latt.
Mellisuga Williami, Gray & Mitch.
Metallura Williami, Bonap., Reichenb.
——— *williami*, Bonap.
*Urolampra Williami, Cab. et Hein. Mus. Hein. Theil iii. p. 68, note.
Habitat. Popayan.

218. METALLURA PRIMOLII, *Bonap.* . . Vol. III. Pl. CXCIV.
Metallura promolina, Bourc., Reichenb.
——— *primolinus*, Bonap.
Urolampra primolina, Cab.
Habitat. Peru.

219. METALLURA TYRIANTHINA . . . Vol. III. Pl. CXCV.
Trochilus tyrianthinus, Lodd.
Ornismya Allardi, Bourc.
——— *Paulinæ*, Boiss.
Trochilus Allardi, Jard.
Mellisuga tyrianthinus, Gray & Mitch.
Metallura tyrianthina, Reichenb.
——— *tyrianthinus*, Bonap.
*Urolampra tyrianthina, Cab. et Hein. Mus. Hein. Theil iii. p. 68.
Habitat. New Granada.

In my account of *Metallura tyrianthina*, I have given that bird a very wide range of habitat, extending from the Gulf of Darien to Ecuador; but having since had ample opportunities for examining numerous specimens from every locality, I find that the birds from Ecuador are so much larger than those from Bogota that I cannot do otherwise than regard them as distinct. In examples from the two localities mentioned, there is a difference of more than half an inch in the length of their wings, and fully an eighth in the length of their bills; I observe also that the small birds from Bogota are much more richly coloured than the larger ones from Ecuador; the throat is of a more beautiful green, the abdomen much darker, and the reddish-purple of the tail more resplendent; believing the Ecuadorian bird to be distinct, I have no alternative but to give it a name, and I therefore propose for it that of *Quitensis*:—

220. METALLURA QUITENSIS, *Gould.*
Habitat. Ecuador.

221. METALLURA SMARAGDINICOLLIS . Vol. III. Pl. CXCVI.
Orthorhynchus smaragdinicollis, D'Orb. et Lafres.
Mellisuga smaragdinicollis, Gray & Mitch.

Metallura smaragdinicollis, Bonap., Reichenb.
* *Urolampra smaragdinicollis*, Cab. et Hein. Mus. Hein. Theil iii. p. 68, note.

Habitat. Peru and Bolivia.

Varied as have been the subjects hitherto referred to in the present volume, and beautiful as is the colouring of many of the species, the next genus is composed of birds which cannot boast of any brilliancy of colouring; on the contrary, they are clothed in very sombre attire, and have nothing to recommend them to our notice but chaste and delicate hues; still in my opinion they are not the less interesting.

Genus ADELOMYIA, *Bonap.*

222. ADELOMYIA INORNATA, *Gould*. . Vol. III. Pl. CXCVII.

Trochilus (——?) *inornata*, Gould.
Mellisuga inornata, Gray & Mich.
Ramphomicron inornatus, Bonap.
Adelomyia inornata, Bonap.
Metallura inornata, Reichenb.
* *Adelisca inornata*, Cab. et Hein. Mus. Hein. Theil iii. p. 72, note.

Habitat. Bolivia.

223. ADELOMYIA MELANOGENYS . . Vol. III. Pl. CXCVIII.

Trochilus melanogenys, Fras.
———— *Sabinæ*, Bourc. et Muls.
Mellisuga Sabinæ, Gray & Mitch.
———— *melanogenys*, Gray & Mitch.
Ramphomicron sabinae, Bonap.
———— *melanogenys*, Bonap.
Adelomyia sabina, Bonap.
Metallura Sabinae, Reich.
* *Adelisca melanogenys*, Cab. et Hein. Mus. Hein. Theil iii. p. 72.

Habitat. New Granada, Ecuador, and Peru.

Precisely the same kind of difference occurs between examples of this form from Venezuela and Ecuador that has been described as occurring with regard to the *Metallura tyrianthina* and *M. Quitensis*. The *Adelomyia* of Ecuador and Peru is very considerably larger than the *A. melanogenys* from Venezuela; it has more buff at the base of the tail-feathers, and a much more conspicuously spotted throat and breast; for this Ecuadorian bird I therefore propose the name of *maculata*:—

224. ADELOMYIA MACULATA, *Gould* . . Vol. III. Pl. CXCIX.

Habitat. Ecuador.

Avocettula and *Avocettinus* are the generic terms applied to the two species rendered remarkable by the points of the mandibles being curved upwards in the shape of a hook: this extraordinary deviation from the usual structure is doubtless designed for some especial purpose; but what that may be, is at present unknown to us.

In placing these two species near to each other I do not mean to convey an idea that they are very nearly allied. One is an inhabitant of the Andes, the other of Guiana and the neighbouring countries. Nothing whatever is known respecting these singular birds.

Genus AVOCETTINUS, *Bonap.*

225. AVOCETTINUS EURYPTERUS . . . . Vol. III. Pl. CC.

Trochilus eurypterus, Lodd.
Polytmus euryptera, Gray & Mitch.
Trochilus Georginæ, Boure.
Polytmus Georginæ, Gray & Mitch.
Delattria georgina, Bonap.
Avocettinus eurypterus, Bonap.
Avocettula euryptera, Reichenb.
——— *Georginae*, Reichenb.
*Opisthoprora euryptera, Cab. et Hein. Mus. Hein. Theil iii. p. 76, note.

Habitat. The high lands of New Granada.

Genus AVOCETTULA, *Reichenb.*

226. AVOCETTULA RECURVIROSTRIS . . . Vol. III. Pl. CCI.

Trochilus recurvirostris, Swains.
Mellisuga? recurvirostris, Steph.
Ornismya recurvirostris, Less.
——— *avocetta*, Less.
Campylopterus recurvirostris, Swains.
Hylocharis recurvirostris, Gray & Mitch.
Avocettinus recurvirostris, Bonap.
——— ——— *lessoni*, Bonap.
Avocettula recurvirostris, Bonap., Reich.
**Trochilus avocetta*, Jard. Nat. Lib. Humming Birds, vol. i. p. 78, pl. 2.
*——— *recurvirostris*, Jard. Nat. Lib. Humming Birds, vol. ii. p. 80.
**Hylocharis avocetta*, Gray & Mitch. Gen. of Birds, vol. i. p. 114, *Hylocharis*, sp. 12.
**Streblorhamphus recurvirostris*, Cab. et Hein. Mus. Hein. Theil iii. p. 76.

Habitat. The Guianas.

Genus ANTHOCEPHALA, *Cab.*

This generic term has been proposed by Dr. Cabanis for the bird I have figured under the name of *Adelomyia floriceps*, which is at present the only species of the form known; for, although I have ventured to place with it my *Adelomyia ? castaneiventris*, I am unable to say, from the imperfect materials at my command, whether it really belongs to the present or to some other genus.

227. ANTHOCEPHALA FLORICEPS.
Adelomyia floriceps, *Gould* . . . . . Vol. III. Pl. CCII.
Trochilus (——?) *floriceps*, Gould.
Adelomyia floriceps, Bonap.
Metallura floriceps, Reichenb.
Anthocephala floriceps, Cab. et Hein. Mus. Hein. Theil iii. p. 72, note.
Habitat. Columbia.

228. ANTHOCEPHALA? CASTANEIVENTRIS.
Adelomyia? castaneiventris, *Gould* . . . Vol. III. Pl. CCIII.
Trochilus (——?) *castaneoventris*, Gould.
Metallura castaneiventris, Reichenb.
Habitat. Chiriqui.

The fourth volume commences with a species which plays no inconsiderable part as an article of trade; for it is the one, *par excellence*, of which thousands are annually sent to Europe for the purpose of contributing to the decorations of the drawing-rooms of the wealthy, for the manufacture of artificial flowers, &c.; and well suited is it for such purposes, its rich ruby and topaz-like colouring rendering it one of the most conspicuous and beautiful objects imaginable. The *Chrysolampis moschitus* (better known by its trivial name of Ruby and Topaz Humming-Bird) enjoys a very wide range, being found all over the eastern parts of Brazil, Cayenne, Guiana, Venezuela, the high lands of Bogota and Trinidad.

The females of this form differ very widely from the males in the colouring of their plumage; and the young males undergo so many changes between youth and maturity, that they must have puzzled the most astute of ornithological investigators.

Genus CHRYSOLAMPIS, *Boié.*

229. CHRYSOLAMPIS MOSCHITUS . . . , Vol. IV. Pl. CCIV.
Trochilus moschitus, Linn. et auct.
Melliruga Brasiliensis, gutture topazino, Briss.
Ornismya moschita, Less.
Chrysolampis moschitus, Boié, Bonap.
Melliruga moschita, Steph., Gray & Mitch.
Chrysolampis mosquitus, Bonap., Reichenb.
Trochilus pegasus, Gmel.
——— *gujanensis*, Gmel., Lath.
——— *carbunculus*, Gmel., Lath.
——— *elatus*, Gmel.
Melliruga Cayanensis, ventre griseo, Briss.
Chrysolampis carbunculus, Reichenb.
Trochilus hypophæus, Shaw.

Chrysolampis moschitus, Cab. et Hein. Mus. Hein. Theil iii. p. 21.
Chrysolampis Reichenbachi, Cab. et Hein. Mus. Hein. Theil iii. p. 21.

Habitat. Guiana, Cayenne, Brazil, Venezuela, the Andes of New Granada, and the islands of Trinidad and Tobago.

Dr. Cabanis is of opinion that the bird from New Granada is distinct from that obtained in the other localities; but I must receive more decided evidence that such is the case than I at present possess, before I can admit that there is any difference between the Andean and Brazilian examples; for the present, therefore, I place his name of *C. Reichenbachi* as a synonym of *C. moschitus*, which I believe to be the only species yet known of the genus.

"This pretty little species," says Mr. Kirk, "arrives in Tobago at the end of January or about the 1st of February. It begins to build about the 10th, lays two pure white eggs, and sits fourteen days. It feeds on ants as well as flowers. I detected 115 small insects in the stomach of one I dissected. One of these birds having attached its nest to the trunk of a logwood tree close to a window of my residence, I had an opportunity of observing its manners during incubation, and I can assert that, although I confined the young by means of some coarse wire cloth, through which the parent could feed them, for upwards of three weeks after they were ready to leave the nest, and although she evinced the greatest distress by her chirping note when flying around me, often within three feet, I never but twice, from the laying until the period I mention, saw a male near the nest; and whether they pair seems to be disputed, as on both these occasions he was hotly pursued by the female to a considerable distance with all the bickering violence so peculiar to the tribe."—*Horæ Zoologicæ*, by Sir W. Jardine, Bart., in *Ann. and Mag. Nat. Hist.* vol. xx. p. 373.

In proceeding next to the genus *Orthorhynchus*, composed of birds ornamented with glittering green and blue crests, I do not insist that they have any direct affinity with the last, nor are they intimately allied to the members of the succeeding one: a more isolated form, in fact, is not to be found among the Trochilidæ. Only two species have been recorded by previous writers, but specimens of a third are contained both in the Loddigesian and my own collections; I allude to the bird here described under the name of *Orthorhynchus ornatus*.

All the members of the

Genus ORTHORHYNCHUS, Cuv.,

are confined to the West India Islands, but our present knowledge of them does not admit of my stating positively the extent of the range of each species; this is a point which requires further investigation. The females differ from the males in being destitute of the glittering crown.

230. ORTHORHYNCHUS CRISTATUS . . . . Vol. IV. Pl. CCV.

Trochilus cristatus, Linn. et auct.

Mellisuga cristata, Briss., Gray & Mitch.
Orthorhynchus cristatus, Bonap., Reichenb.
Ornismya cristatus, Less.
Trochilus pileatus, Lath.
———— *puniceus*, Gmel.
* *Orthorhynchus cristatus*, Cab. Mus. Hein. Theil ill. p. 61.

Habitat. Barbadoes, and St. Vincent.

The Rev. Lansdown Guilding states that this species "sometimes deviates from its usual habits. In general it is remarkably wild, and soon disturbed. I once, however, saw a pair of this species almost domesticated, in the house of a gentleman whose kindness and humanity had brought round him many a lizard and winged pet. They built for many years on the chain of the lamp suspended over the dinner-table; and here they educated several broods, in a room occupied hourly by the family. I have been seated with a large party at the table when the parent bird has entered, and, passing along the faces of the visitors, displaying his glorious crest, has ascended to the young without alarm or molestation."—*London's Mag. Nat. Hist.* vol. v. p. 670.

231. ORTHORHYNCHUS ORNATUS, *Gould* . . Vol. IV. Pl. CCVI.
* *L'Oiseau-mouche huppé*, Less. Hist. Nat. des Ois.-mou. p. 113, pls. 31, 52?

Habitat. One of the Windward Islands; but which of them, is unknown.

Basal two-thirds of the crest green, the apical third blue; upper surface deep grass green; throat dark smoky grey, becoming much darker on the abdomen: flanks glossed with green; wings and tail purplish black; bill black.
Total length $3\frac{1}{4}$ inches; bill, $\frac{1}{2}$; wing $1\frac{7}{8}$; tail $1\frac{1}{4}$.

This species bears a general resemblance to the *O. cristatus*, but differs from that bird in being of a somewhat smaller size, and in having the basal two-thirds of the crest glittering green and the tip only blue; the crest is also longer and more elegant in form than that of *C. cristatus* or *C. exilis*. With the latter it never can be confounded, while the former may always be distinguished from it by the truncate form of the green portion of its crest. It is just possible that the birds represented on the 31st and 32nd plates of Lesson's 'Histoire Naturelle des Oiseaux-mouches' may have reference to this bird.

232. ORTHORHYNCHUS EXILIS . . . . . Vol. IV. Pl. CCVII.
Trochilus exilis, Gmel., Lath., Vieill.
Mellisuga exilis, Gray & Mitch.
Trochilus cristatellus, Lath., Vieill.
Orthorhynchus chlorolophus, Bonap.
———— *exilis*, Reichenb., Bonap., Cabanis.

Habitat. The Islands of Martinique, Nevis, St. Thomas, and St. Croix.

"After a careful examination of skins procured from St. Croix and St. Thomas," says Mr. Alfred Newton, "we refer them to the above-named species, though one of a male presents a slightly different appearance from the ordinary type, in having a narrow blue edging to the otherwise golden-green crest, and thus exhibiting an affinity to the closely-allied Blue-crest (*O. cristatus*) from St. Vincent and Barbadoes. The present bird has, we believe, hitherto been known only from Martinique and Nevis.

"I shot a female of this species at Southgate Farm on the north shore of the eastern end of the Island of St. Croix, where much of the land, being out of cultivation, is chiefly covered with Casha bushes, interspersed with Manchioneel along the coast. I have been told that a Humming-Bird smaller than the ordinary one, and therefore probably of this species, has been seen in other localities; but it must be very uncommon. Of its habits I know nothing."—*Ibis*, vol. i. p. 141.

The Brazilian genus *Cephalepis* comprises two species, with lengthened ornamental crests terminating in a single plume, on which account they stand alone not only in their own family, but, so far as I am aware, among birds generally. The females are entirely devoid of this conspicuous character. I think it very probable that additional species of this form will be discovered when the natural productions of the interior of Brazil become better known.

Genus CEPHALEPIS, *Boié*.

233. CEPHALEPIS DELALANDI . . . . Vol. IV. Pl. CCVIII.

Trochilus Delalandi, Vieill., Temm., Valenc., Less. &c.
Ornismya Delalandi, Less.
Trochilus verricolor, Vieill.
Mellisuga Delalandi, Gray & Mitch.
Cephalepis lalandii, Bonap.
* *Cephalepis Delalandii*, Reichenb. Aufz. der Col. p. 12.
* *Orthorhynchus Delalandii*, Reichenb. Troch. Enum. p. 9.
* *Cephalolepis Delalandi*, Cab. et Hein. Mus. Hein. Theil iii. p. 61.

Habitat. Southern Brazil.

234. CEPHALEPIS LODDIGESI, *Gould* . . . Vol. IV. Pl. CCIX.

Trochilus Loddigesii, Gould, Less., Jard.
———— *opisthoconus*, Licht.
Cephalepis loddigesi, Bonap.
Mellisuga Loddigesii, Gray & Mitch.
* *Cephalepis Loddiggesii*, Reichenb. Aufz. der Col. p. 12.
* *Orthorhynchus Loddiggesii*, Reichenb. Troch. Enum. p. 9.
* *Cephalolepis Loddigesi*, Cab. et Hein. Mus. Hein. Theil iii. p. 61, note.

Habitat. Minas Geraes and other parts of eastern Brazil.

Near to *Cephalepis* is the

Genus KLAIS, *Reichenb.*,

of which but one species is known. This singular bird, which has no ornamental crest, and but little fine colouring to recommend it to our notice, is a native of Venezuela and the hilly parts of New Granada. The females of this form are much less highly coloured than the males.

235. KLAIS GUIMETI . . . . Vol. IV. Pl. CCX.

Trochilus Guimeti, Bourc et Muls.
Hylocharis Guimeti, Gray & Mitch.
Klais Guimeti, Reichenb.
Myiabeillia guimeti, Bonap.
* *Basilinna Guimeti*, Reichenb. Troch. Enum. p. 12; Cab. et Hein. Mus. Hein. Theil iii. p. 45.
* *Myiabellia guimeti*, Sclat. in Proc. Zool. Soc. part xxv. p. 17.
* *Mellisuga Merrittii*, Lawr. Ann. Lyc. Nat. Hist. in New York, vol. vii. April 9, 1860.

Habitat. Venezuela and the Andes of New Granada.

"In the district of El Mineral" in New Granada, says Dr. Merritt, " there has been a slight attempt at cultivation of the soil and planting of fruit trees. The Orange, the Guama, and Guayava trees are the most numerous, particularly the last named, which is very prolific, bearing nearly throughout the year fruit in all its stages from the blossom to maturity. Consequently the Guayava tree is the favourite resort of the Humming Bird. I often watched these little creatures feeding and quarrelling around a tree near the door of my palm-leaf hut, and soon my attention was attracted to one much smaller than the rest, whose pugnacity and indomitable ' pluck' greatly amused me. Upon closer examination of this diminutive feathered warrior my interest increased, as I soon became convinced that it was new to me. I frequently afterwards saw numerous specimens of it, and almost invariably encountered them feeding from the blossoms of the Guayava, and I therefore conclude they are quite local in their habitat."

Genus MYIABEILLIA, *Bonap.*

This is a very distinct generic form. Like that of *Klais*, it contains but a single species, a delicately formed bird inhabiting Mexico and Guatemala. The male is decorated with a brilliant green gorget, a feature which is entirely absent in the female.

236. MYIABEILLIA TYPICA . . . . . . Vol. IV. Pl. CCXI.

Trochilus Abeillei, Delatt. et Less.
Mellisuga Abeillei, Gray & Mitch.
Ramphomicron abeillei, Bonap.
Myiabeillia typica, Bonap.
* *Bouers Abeillei*, Reichenb. Aufz. der Col. p. 13.
* *Abeillia typica*, Bonap. Consp. Gen. Av. tom. i. p. 79, *Ramphomicron*, sp. 4.

* *Barilinna Abeillii*, Reichenb. Troch. Enum. p. 11.
* *Baucis Abeillei*, Cab. et Hein. Mus. Hein. Theil III. p. 72.
Habitat. Mexico and Guatemala.

"Volcan de Fuego and Coban. The barrancos of the volcano are the only localities I am aware of, near Dueñas, where this species is found. Here, however, it is a common bird. It is usually to be seen feeding about upon the brushwood, seeking the flowers, &c. It is a restless species, but shows little symptoms of fear. My skins from the Volcano are one female and three males. The proportions at Coban are very different. Here it is common, being found in all the mountain-hollows feeding among the *Salviæ*. The ratio of the sexes is as twenty males to one female."—*Salvin in Ibis*, vol. ii. p. 262.

I must now direct the notice of my readers to some of the most chaste and elegant species yet discovered of this or any other family of birds—namely, the members of the genera *Heliactin*, *Heliothrix*, *Schistes*, and *Petasophora*.

Genus HELIACTIN, *Boié*.

The single species of this genus stands alone for the resplendent and richly coloured tufts of feathers which spring from above and behind the eye.

237. HELIACTIN CORNUTA . . . . . . Vol. IV. Pl. CCXII.

Trochilus cornutus, Pr. Max., Jard.
———— *dilophus*, Vieill.
———— *bilophus*, Temm.
Ornismya chrysolopha, Less.
Melliruga bilopha, Steph.
Heliactin cornuta, Bonap.
Mellisuga cornuta, Gray & Mitch.
Heliactinia cornuta, Reichenb.
* *Heliactinus cornutus*, Burm. Th. Bras. tom. ii. p. 356.
* *Trochilus bilobus*, Licht. in Mus. of Berlin.
* *Heliactin cornuta*, Cab. et Hein. Mus. Hein. Theil III. p. 64.

Habitat. Brazil, particularly the district of Minas Geraes.

The female of this species, like that sex in the genus *Heliothrix*, has a longer and more ample tail than her mate,—in which respect the form offers an affinity to the members of the succeeding genus.

Genus HELIOTHRIX, *Boié*.

This, like the last, is a very well-marked form, of which two species are natives of Central America and New Granada, one of the regions of the upper Rio Negro, and two of the rich country of Brazil. They are all distinguished by being decorated with beautiful blue tufts on the sides of the neck, relieved by glittering green cheeks and snowy breasts. In addition to this fine display of colours, two of them have rich blue crowns; there is yet another character common, I believe, to the whole—that of the tail of the females being much larger and more lengthened than that of the males; the young males, too, have this organ much more prolonged than in the adult males; they have all peculiarly sharp wedge-shaped bills, lengthened wings, and small feet. Judging from these points in their structure, I believe these birds to be endowed with the power of more rapid flight than any other members of the family.

238. HELIOTHRIX AURITUS . . . . . Vol. IV. Pl. CCXIII.
    Trochilus auritus, Gmel., Lath., Vieill., Swains.
    Mellisuga Cayenensis major, Briss.
    Ornismya aurita, Less.
    Heliothrix auritus, Boié, Gray & Mitch., Bonap.
    ————— *aurita*, Gray.
    Ornismya nigrotis, Less.
    Heliothrix nigrotis, Gray & Mitch., Bonap.
    * *Trochilus leucocrotaphus*, Vieill. Nouv. Dict. d'Hist. Nat. tom. vii. p. 374; Id. Ency. Méth. Orn. part. 2nde, p. 571.
    * *Heliothrix aurita*, Reichenb. Aufz. der Col. p. 13; Troch. Enum. p. 11; Cab. et Hein. Mus. Hein. Theil iii. p. 28.
    Habitat. Northern Brazil, the banks of the Amazon, the Guianas, and Venezuela.

239. HELIOTHRIX AURICULATUS . . . . Vol. IV. Pl. CCXIV.
    Trochilus auriculatus, Licht.
    Ornismya Pouchettii, Less.
    Heliothrix auriculatus, Gray & Mitch.
    ————— *poucheti*, Bonap.
    * *Heliothrix auriculata*, Reichenb. Aufz. der Col. p. 13; Id. Troch. Enum. p. 11; Cab. et Hein. Mus. Hein. Theil iii. p. 28.
    * ————— *aurita, jun.*, Burm. Th. Bras. tom. ii. p. 336.
    Habitat. South-eastern Brazil.

240. HELIOTHRIX PHAINOLÆMA, *Gould* . Vol. IV. Pl. CCXV.
    * *Heliothrix phænoleuca*, Hartl. Wieg. Arch. xxii. 2. p. 29.
    ————— *phænolæma*, Cab. et Hein. Mus. Hein. Theil iii. p. 28, note.
    Habitat. The banks of the Upper Rio Negro.

241. HELIOTHRIX BARROTI.
    Heliothrix purpureiceps, *Gould* . . . Vol. IV. Pl. CCXVI.
    Heliothrix purpuriceps, Gould in Proc. Zool. Soc., part xxiii. p. 87.
    ————— *Barroti*, Salvin in Ibis, vol. iii. p. 410.
    Trochilus Barroti, Boore.
    Heliothrix Barroti, Gray & Mitch., Bonap.
    * *Heliothrix Barroti*, Reichenb. Aufz. der Col. p. 13; Id. Troch. Enum. p. 11; Cab. et Hein. Mus. Hein. Theil iii. p. 28 note.

Ornismya Gabriel, Delatt. Echo du Monde Savant, No. 45, Juin 15, 1843, col. 1070.

Habitat. Guatemala, Costa Rica, and the forests of New Granada bordering the Pacific coast, as far south as Ecuador; Esmeraldas (*Fraser*).

242. HELIOTHRIX VIOLIFRONS, *Gould.*

Heliothrix Barroti . . . . Vol. IV. Pl. CCXVII.

Habitat. Carthagena, or Veragua.

On receiving this bird from M. Warszewicz, I considered it to be referable to the *Heliothrix Barroti*, and accordingly figured and described it under that name. Subsequently I received another bird with a differently coloured crown, which, believing it to be new, I described and figured as *H. purpureiceps*, but I now find that the latter is the true *H. Barroti*, and that the former is a new bird; I therefore propose for it the distinctive appellation of *H. violifrons*.

Between the genera *Heliothrix* and *Petasophora* appears to be the proper situation for my genus *Schistes*; for to the former it is nearly allied in its wedge-shaped bill, and to the latter in the colouring of the tail. The three species known are all inhabitants of the Andes of New Granada and Ecuador. I have often thought that the white gular mark in *Schistes albigularis* is characteristic of immaturity; but this is by no means certain.

Genus SCHISTES, *Gould.*

(Σχίζω, findo.)

Generic characters.

Male.—*Bill* longer than the head, straight, wedge-shaped at the tip; *wings* moderately long and slightly rounded; *tail* rounded, the feathers broad; *tarsi* partially clothed; *feet* small; *hind toe* and *nail* shorter than the middle toe and nail.

243. SCHISTES GEOFFROYI . . . . . Vol. IV. Pl. CCXVIII.

Trochilus Geoffroyi, Bourc. et Muls.
Petasophora? Geoffroyi, Gould.
Polytmus Geoffroyi, Gray & Mitch.
Colibri Geoffroyi, Bonap.
*Schistes Geoffroyi, Reichenb. Aufz. der Col. p. 13.
*Petasophora Geoffroyi, Reichenb. Troch. Enum. p. 11.
*Schistes Geoffroyi, Cab. et Hein. Mus. Hein. Theil iii. p. 27.

Habitat. The Andes of New Granada.

244. SCHISTES PERSONATUS, *Gould* . . Vol. IV. Pl. CCXIX.

Schistes geoffroyii, Sclat. in Proc. Zool. Soc. part xxviii. p. 70.

Habitat. Ecuador.

This bird was obtained by Mr. Fraser at Pallatanga. He remarks, "I should have taken this to be the male of *S. albogularis* but for the colour of the feet and shape of the tail. Bill and feet black. Stomach contained insects."

245. SCHISTES ALBIGULARIS, *Gould* . . Vol. IV. Pl. CCXX.
Schistes albigularis, Reichenb. Aufz. der Col. p. 13.
Petasophora albigularis, Ib. Troch. Enum. p. 11.
Schistes albigularis, Cab. et Hein. Mus. Hein. Theil iii. p. 27, note.
Habitat. The western side of Pichincha, in Ecuador, at an elevation of 6000 feet.

Mr. Fraser, who procured this bird at Pallatanga, says, " Bill black; feet dark flesh-colour; gizzard contained insects; found in the underwood."

The members of the next genus, *Augastes*, have perhaps no direct alliance with the preceding; but as they are characterized by masked faces, and have buffy marks on the sides of the chest, they are as well placed here as elsewhere.

Both the *A. scutatus* and *A. Lumachellus* are very beautiful species, and have had the trivial name of Vizor-bearers applied to them, from the very peculiar manner in which their entire faces are covered with shining metallic feathers, giving the birds the appearance of being masked; the under surface of their tails is also luminous, in which respect they present a similarity to the *Metallurœ*.

Genus AUGASTES, *Gould*.

(Αὐγάζω, illucesco, de αὐγή, splendor.)

Male.—Bill straight, longer than the head, and inclining to a wedge-shape at the tip; *head* round, the feathers not advancing upon the bill; *wings* rather long; *tail* moderately long and square, the feathers broad; *tarsi* clothed; *feet* small; *hind toe* very diminutive; *face* and tail luminous.
Female.—Destitute of luminous colouring.

246. AUGASTES SCUTATUS . . . . . Vol. IV. Pl. CCXXI.
Trochilus superbus, Vieill.
――――― *scutatus*, Natt., Temm., Jard.
Ornismya Nattererii, Less.
Hylocharis superba, Gray & Mitch.
Trochilus venustus, Licht. in Mus. of Berlin.
Augastes superbus, Reichenb. Aufz. der Col. p. 13; Troch. Enum. p. 11; Bonap. Rev. et Mag. de Zool. 1854, p. 253.
Habitat. Ecuador.

247. AUGASTES LUMACHELLUS . . . Vol. IV. Pl. CCXXII.
Ornismya lumachella, Less.
Trochilus Lumachellus, Bourc.
Hylocharis Lumachellus, Gray & Mitch.
Lamprurus Lumachellus, Reichenb. Aufz. der Col. p. 12.
Ramphomicron Lumachellus, Reichenb. Troch. Enum. p. 10.
Augastes lumachellus, Bonap. Rev. et Mag. de Zool. 1854, p. 253; Cab. et Hein. Mus. Hein. Theil iii. p. 46.
Habitat. Central and Northern Brazil.

One of those genera which give but little trouble to the ornithologist is the

Genus PETASOPHORA, *G. R. Gray*,

all the species having characters in common, while each has its own peculiar distinction either in colour or markings. The sexes are alike in colour, but the females are always much smaller than the males. This is strictly an Andean group, most of the species being found in those elevated regions from Mexico in the north to Bolivia in the south; one species, the *P. serrirostris*, inhabits Brazil.

248. PETASOPHORA SERRIROSTRIS . . Vol. IV. Pl. CCXXIII.

Trochilus serrirostris, Vieill.
——— *janthinotus*, Natt.
——— *petasophorus*, Pr. Max., Temm., Jard.
Ornismya petasophorus, Less.
Grypus? Vieilloti, Steph.
Colibri crispus, Spix.
Petasophora serrirostris, Gray, Gould, Bonap.
Polytmus serrirostris, Gray & Mitch.
* *Trochilus (Lophornis) petasophorus*, Tschudi, Consp. p. 97. No. 205.
* ——— *chalcotis*, Licht. in Mus. of Berlin.
* *Petasophora chalcotis*, Reichenb. Aufz. der Col. p. 13; Id. Troch. Enum. p. 11.
* ——— *serrirostris*, Id. ib. p. 13.
* ——— *crispa*, Burm. Th. Bras. tom. ii. p. 335.
* ——— *serrirostris*, Cab. et Hein. Mus. Hein. Theil iii. p. 25.
Petasophora Gouldi, Bonap. (proposed for a smaller bird inhabiting Bahia).

Habitat. Brazil, from Minas Geraes to Bahia.

249. PETASOPHORA ANAIS . . . . . . Vol. IV. Pl. CCXXIV.

Ramphodon Anais, Less.
Polytmus Anais, Gray & Mitch.
Colibri anais, Bonap.
Trochilus thalassinus, Jard.
——— *Anais*, Jard.
* *Petasophora Anais*, Bonap. Rev. et Mag. de Zool. 1854, p. 250; Reichenb. Troch. Enum. p. 11; Cab. et Hein. Mus. Hein. Theil iii. p. 26.
* *Praxilla Anais*, Reichenb. Aufz. der Col. p. 13.

Habitat. New Granada and Venezuela.

250. PETASOPHORA IOLATA, *Gould* . . Vol. IV. Pl. CCXXV.

Polytmus iolatus, Gray & Mitch.
Colibri iolata, Bonap.
*? *Trochilus (Coeligena) Anais*, Tschudi, Consp. p. 36, No. 201; Id. Faun. Peru. p. 244, No. 4.
* *Praxilla iolata*, Reichenb. Aufz. der Col. p. 13.
* *Petasophora iolata*, Reichenb. Troch. Enum. p. 11.

* *Petasophora rhodotis* " Gould," Saucerotte in Mus. Heinean.
* ——— *iolata*, Cab. et Hein. Mus. Hein. Theil iii. p. 26.
Habitat. Ecuador, Peru, and Bolivia.

251. PETASOPHORA CORUSCANS, *Gould* . Vol. IV. Pl. CCXXVI.

Trochilus (Petasophora) coruscans, Gould.
Petasophora coruscans, Gould.
Polytmus coruscans, Gray & Mitch.
Colibri coruscans, Bonap.
Habitat. Unknown.

I have never seen a second example of this singular bird, which departs from the ordinary species, and assimilates somewhat to the *P. Delphinæ*.

252. PETASOPHORA THALASSINA . . Vol. IV. Pl. CCXXVII.

Trochilus thalassinus, Swains.
Polytmus thalassinus, Gray & Mitch.
Colibri thalassinus, Bonap.
* *Trochilus Anais*, Swains. Birds of Brazil, pl. 75.
* *Ornismya Anais*, Less. Supp. des Ois.-mou. pl. 32.
* *Ramphodon Anais*, Less. Troch. p. 148, pl. 56 ?
* *Trochilus Annis*, Jard. Nat. Lib. Humming Birds, vol. ii. p. 2?
* *Cynanthus thalassinus*, Jard. ib. p. 148.
* *Colibris thalassina*, Sclat. in Proc. of Zool. Soc. part. xxiv. p. 287.
* *Petasophora thalassina*, Bonap. Rev. et Mag. de Zool. 1854, p. 250; Cab. et Hein. Mus. Hein. Theil iii. p. 27; Reichenb. Troch. Enum. p. 11.
* *Prurilla thalassina*, Reichenb. Aufz. der Col. p. 13.

Habitat. Mexico and Guatemala.

" The barrancos of the Volcan de Fuego are favourite resorts of this species. A specimen obtained at Dueñas on the 15th of September was the only one I saw out on the llano, as the bird is usually found in the dense forest."—*Salvin* in *Ibis*, vol. ii. p. 260.

253. PETASOPHORA CYANOTIS . . . Vol. IV. Pl. CCXXVIII.

Trochilus cyanotus, Bourc.
Petasophora cyanotus, Gould.
Polytmus cyanotus, Gray & Mitch.
Colibri cyanotis, Bonap.
* *Ornismya Anais*, Less. Troch. p. 151, pl. 57 ?
* *Prurilla cyanotis*, Reichenb. Aufz. der Col. p. 13.
* *Petasophora cyanotus*, Reichenb. Troch. Enum. p. 11.
* ——— *cyanotis*, Bonap. Rev. et Mag. de Zool. 1854, p. 251; Cab. et Hein. Mus. Hein. Theil iii. p. 26.

Habitat. New Granada and Venezuela.

254. PETASOPHORA DELPHINÆ . . . Vol. IV. Pl. CCXXIX.

Ornismya Delphinæ, Less.
Polytmus Delphinæ, Gray & Mitch.

Colibri delphinæ, Bonap.
* *Telesiella Delphinæ*, Reichenb. Aufz. der Col. p. 19.
* *Petasophora Delphinæ*, Reichenb. Troch. Enum. p. 11.
* ———— *delphina*, Bonap. Rev. et Mag. de Zool. 1854, p. 251.
* *Tilesilla Delphinæ*, Cab. et Hein. Mus. Hein. Theil iii. p. 27.

Habitat. The Guianas, Trinidad, Venezuela, Guatemala, New Granada, and Ecuador.

"This Humming-Bird seems to have been quite unknown at Coban previously to the collection of my specimens. The first was shot by my collector, Cipriano Prado, among some *Salviæ*, in one of the mountain hollows near Coban. *Salviæ* being in flower in November, their blossoms are sought after by nearly every species of Humming-Bird near Coban, and this among the rest. It is rare even at Coban; and though much sought for by the Indian boys in consequence of my offers of reward, but few specimens were obtained.

"Three males to one female appears to be about the proportion of the sexes."—*Salvin* in *Ibis*, vol. ii. p. 261.

There is no one genus among the Trochilidæ that has more sadly puzzled me, and doubtless other ornithologists, than that containing the two species known under the specific names of *virescens* and *viridissimus* (*Chrysobronchus virescens* and *C. viridicaudus* of my Plates), no two persons agreeing as to the place they should fill in the family. Dr. Cabanis, in his 'Museum Heineanum,' is of opinion that the generic name of *Polytmus*, proposed by Brisson in 1760, is the one under which they should be retained; although I concur in this opinion, I cannot agree with him in placing them near to the genus *Glaucis*; and I may be open to criticism in ranging them here, but I really cannot find a better situation for them. I have stated that there are two species of this form, but I have some reason to believe there is a third, as I have a small specimen collected by M. Warszewicz on the River Magdalena, which may prove to be distinct; but until I have further evidence that such is the case, I decline to characterize it; independently of its smaller size, it has much more white on the tail than any other I have seen.

Genus POLYTMUS, *Briss.*

The *P. virescens* and *P. viridissimus* are the only species yet characterized of this genus. They are distinguished by the golden hues of their throats. A great similarity exists between the sexes; but the young of *P. virescens* have reddish-brown breasts, and are altogether different in colour from the adults.

255. POLYTMUS VIRESCENS.

Chrysobronchus virescens . . . . . Vol. IV. Pl. CCXXX.
Trochilus Thaumatias, Linn., Lath., Vieill.
———— viridescens, Linn.
———— virescens, Dumont, Licht., Vieill., Pr. Max.
———— chrysobronchus, Shaw, Steph.

Trochilus viridis, Vieill.
Ornismya viridis, Less.
Trochilus chloroleucurus, Sauc.
Polytmus chrysobronchus, Gray & Mitch.
Chrysobronchus virescens, Bonap.
Leucippus chrysobronchus, Reichenb.
* *Polytmus thaumantias*, Cab. et Hein. Mus. Hein. Theil iii. p. 5.

Habitat. Trinidad; Venezuela; and New Granada?

256. POLYTMUS VIRIDISSIMUS.
   Chrysobronchus viridicaudus . . . Vol. IV. Pl. CCXXXI.
Trochilus viridissimus, Aud. et Vieill.
Trochilus Theresiæ, Da Silva.
* *Ornismya viridis*, Less. Les Troch. p. 96, pl. 93.
* *Trochilus virescens*, Wied, Beitr. iv. p. 107.
* *Amazilia viridissima*, Bonap. Gen. Av. tom. i. p. 77, *Amazilia*, sp. 4.
* *Smaragditis viridissima*, Reichenb. Aufz. der Col. p. 7.
* *Chrysobronchus viridissimus*, Bonap. Rev. et Mag. de Zool. 1854, p. 252.
* *Chlorestes viridissimus*, Reichenb. Troch. Enum. p. 4, pl. 695. figs. 4547-48.
* *Thaumatias viridissimus*, Burm. Th. Bras. tom. ii. p. 344.
* *Thaumatias chrysurus*, Burm. ib. p. 345.
* *Trochilus viridicaudus*, Sauc. MSS.
* *Trochilus prasinus*, Licht. in Mus. of Berlin.
* *Polytmus Theresiæ*, Cab. et Hein. Mus. Hein. Theil iii. p. 5.

Habitat. The banks of the Amazon, from Para to the confines of Peru. "I have also received specimens from Demerara.

I have before stated that it would be impossible to arrange the Humming-Birds on the score of affinity; and I repeat that the various genera are so widely different, and so many connecting forms are wanting, that it is quite out of the question to attempt their arrangement on this ground. It is of little importance, then, where we place the bird known under the name of *Patagona gigas* and distinguished from all others by its great size, its ample wings, its sombre colouring, and by the similarity in the plumage of the two sexes. At present the single species which has been characterized, and which ranges from Ecuador to the southern parts of Chili, where it is a migrant, is all that is known of this form; but I observe that the Chilian and Ecuadorian specimens differ considerably in size, the latter being the largest.

Genus PATAGONA, *G. R. Gray.*

257. PATAGONA GIGAS . . . . . Vol. IV. Pl. CCXXXII.
Trochilus gigas, Vieill., Jard.
Ornismya tristis, Less.

Cynanthus tristis, Less.
Ornismya gigantea, D'Orb. et Lafres.
Patagona gigas, Gray, Bonap., Reichenb.
Hylocharis gigas, Gray.
*Hypermetra gigas, Cab. et Hein. Mus. Hein. Theil iii. p. 81.
*Trochilus gigas, Bridges, Proc. of Zool. Soc. part xi. p. 114;
Darwin, Zool. of Beagle, part iii. Birds, p. 111.

Habitat. Ecuador, Peru, Bolivia, and Chile.

"The American Aloe (*Agave Americana*) is the only plant this bird is ever seen feeding upon in Ecuador."—*Jameson and Fraser* in *Ibis*, vol. i. p. 400.

"This species," says Mr. Darwin, "is common in Central Chile. It is a large bird for the delicate family to which it belongs. At Valparaiso, in the year 1834, I saw several of these birds in the middle of August, and I was informed they had only lately arrived from the parched deserts of the north. Towards the middle of September (the vernal equinox) their numbers were greatly increased. They breed in Central Chile, and replace, as I have before said, the foregoing species" (*Eustephanus galeritus*), "which migrates southward for the same purpose. The nest is deep in proportion to its width—externally three inches and a half deep, internal depth a little under one inch and three quarters, width within one inch and two-tenths; mouth slightly contracted. Externally it is formed of fine fibrous grass woven together, and attached by one side, and bottom, to some thin upright twigs; internally it is thickly lined with a felt, formed of the pappus of some composite flower. When on the wing, the appearance of this bird is singular. Like others of the genus, it moves from place to place with a rapidity which may be compared to that of *Syrphus* among Diptera, and *Sphinx* among Moths; but whilst hovering over a flower, it flaps its wings with a very slow and powerful movement, totally different from that vibratory one, common to most of the species, which produces the humming noise. I never saw any other bird, where the force of its wings appeared (as in a butterfly) so powerful in proportion to the weight of its body. When hovering by a flower, its tail is constantly expanded and shut like a fan, the body being kept in a nearly vertical position. This action appears to steady and support the bird, between the slow movements of its wings. Although flying from flower to flower in search of food, its stomach generally contained abundant remains of insects, which I suspect are much more the object of its search than honey is. The note of this species, like that of nearly the whole family, is extremely shrill."—*Darwin, Zoology of the Beagle*, part iii. Birds, p. 111.

"The *Troch. gigas* is found in all the central provinces of Chile; it is seen about Valparaiso during the spring and summer months, feeding on the flowers of *Pourretia coarctata* and *Lobelia polyphylla* in preference to others. It generally builds its nest near a little rivulet, frequently on a solitary twig or branch over the water; the nest is beautifully constructed, and is composed of moss and the down of a species of *Gnaphalium*. Eggs white; iris dark brown. Catches flies."—*Bridges* in *Proc. Zool. Soc.* pt. vi. p. 114.

The forms to which we now proceed are mostly of large size, have straight lengthened bills, and are very gorgeously coloured. These straight and prolonged bills are in unison with the flora with which they are associated, particularly such deep tubular flowers as those of the genera *Brugmansia*, *Lepageria*, *Nematanthus*, *Tacsonia*, *Alstræmeria*, *Dipladenia*, &c.

The first genus is that of *Docimastes*. Of this remarkable form, the single species known stands alone among Humming-Birds for the great length of its bill. Nature here appears to have carried the development of this organ to its maximum; and how wonderfully is it adapted for exploring the lengthened tubular flowers from which the bird obtains its insect food!

Genus DOCIMASTES, *Gould*.

(Δοκιμάζω, exploro.)

Generic characters.

Male.—*Bill* of extraordinary length, exceeding that of the head and body, and inclining upwards; *wings* long and pointed; *tail* moderately long and forked; *tarsi* short and partially clothed; *feet* small; *hind toe* shorter than the middle toe; *face* and *gorget* dull; *sides of the chest* luminous.

Female.—Unadorned.

258. DOCIMASTES ENSIFERUS . . . Vol. IV. Pl. CCXXXIII.

Ornismya ensifera, Bois.
Trochilus Derbianus, Fras.
Mellisuga ensifera, Gray & Mitch.
* *Docimastes Derbyanus*, Licht. in Mus. Berlin.
* ————— *ensifera*, Cab. et Hein. Mus. Hein. Theil iii. p. 77.

Habitat. Columbia and Ecuador.

Specimens from the neighbourhood of Bogota differ from those received from Quito in being of much smaller size; but I consider these as mere races of one and the same species, for I can see no characters on which a specific distinction could be founded.

The next species is interesting for its great size, the elegance of its proportions, and the beauty and harmony of its colours. This new and extraordinary bird I have named *Eugenia Imperatrix* in honour of the Empress of the French.

Genus EUGENIA, *Gould*.

(*Eugenia*, nom. propr.)

Generic characters.

Male.—*Bill* straight or slightly inclining upwards, longer than the head; *wings* long; *primaries* rigid; *tail* long and forked, the feathers narrow and unyielding; *tarsi* partially clothed; *feet* small; *hind toe* long; *nail* moderate; *face* luminous. *Female.*—Unadorned.

259. EUGENIA IMPERATRIX . . . . Vol. IV. Pl. CCXXXIV.

Habitat. Ecuador.

"Professor Jameson's specimens of this fine bird were obtained in the neighbourhood of Auca, on the road to Nanegal, at about 6000 or 7000 feet elevation. They were feeding on the *Alstræmeriæ, Daturæ* not being found in that locality."—*Jameson and Fraser* in *Ibis,* vol. i. p. 400.

The members of the genus *Helianthea,* distinguished by their starlike frontlets and luminous under surfaces, appear to range next to the preceding. Three of them (namely, *H. typica, H. Bonapartei,* and *H. Eos*) are quite typical; while the *H. Lutetiæ* and *H. violifera* differ somewhat in their colouring, the lower part of the body of the two latter species not being luminous, while they assimilate in all other respects. Dr. Reichenbach's separation of the *H. typica* and *H. Bonapartei* into a separate genus (*Hypochrysia*) cannot, in my opinion, for a moment be admitted.

Genus HELIANTHEA, *Gould.*

("Ἥλιος, sol, et ἄνθος, flos.)

Generic characters.

Male.—*Bill* long, straight and cylindrical; *wings* moderately long and powerful; *tail* of medium size and slightly forked when closed; *tarsi* extremely short and clothed with feathers; *feet* very small; *hind toe* the shortest; *forehead* and *under surface* luminous.

Female.—Destitute of luminous colouring.

The members of this genus frequent the Andes for at least eight degrees on each side of the equator.

260. HELIANTHEA TYPICA . . . . Vol. IV. Pl. CCXXXV.

Ornismya helianthea, Less.
Mellisuga helianthea, Gray & Mitch.
Helianthea typica, Bonap., Cabanis.
**Trochilus porphyrogaster,* Licht. in Mus. of Berlin.

Habitat. New Granada. Is exceedingly common in the neighbourhood of Bogota. A large race occurs near Pamplona.

261. HELIANTHEA BONAPARTEI . . Vol. IV. Pl. CCXXXVI.

Ornismya Bonapartei, Boiss., Bourc.
Trochilus aurigaster, Lodd.
Mellisuga Bonapartei, Gray & Mitch.
Helianthea Bonapartii, Bonap., Cabanis.
**Hypochrysa Bonaparti,* Reichenb. Aufz. der Col. p. 9; Id. Troch. Enum. p. 6, pl. 739. figs. 4683-84.
**Trochilus chrysogaster,* Licht. in Mus. Berlin.

Habitat. New Granada. Examples frequently occur in collections from Bogota.

262. HELIANTHEA EOS, *Gould* . . Vol. IV. Pl. CCXXXVII.
Mellisuga eos, Gray & Mitch.
Helianthea eos, Bonap.
Hypochrysia eos, Reichenb.
Habitat. Paramos de los Conejos, near Merida in Columbia.

263. HELIANTHEA LUTETIÆ . . . Vol. IV. Pl. CCXXXVIII.
Trochilus Lutetiæ, Delatt. et Bourc.
Mellisuga lutetiæ, Gray & Mitch.
Helianthea lutetiæ, Bonap., Cabanis.
Habitat. Popayan and Ecuador. Professor Jameson and Mr. Fraser state that "This bird is found in the valleys of Lloa and Pelogalli, but not nearer Quito."—*Ibis*, vol. i. p. 400.

264. HELIANTHEA VIOLIFERA, *Gould* Vol. IV. Pl. CCXXXIX.
Trochilus violifer, Gould.
Mellisuga violifera, Gray & Mitch.
Helianthea violifera, Bonap.
——— *violifera*, Bonap., Cabanis.
Habitat. "In provinz Chulimani au Cordilera" in Bolivia (*Warszewicz*).

Genus HELIOTRYPHA, *Gould.*

("Ἥλιος, sol, et τρυφή, luxuria.)
Generic characters.
Male.—Bill straight and of the same length as the head; *wings* rather long; *tail* long and forked; *tarsi* partially clothed; *feet* small; *hind toe* rather shorter than the middle one; *forehead* and *throat* luminous.
Female.—Destitute of luminous colouring on the throat.
The members of this genus, two in number, differ from those of *Heliangelus* in the absence of any band of white on the chest and in having a lengthened and deeply forked tail.

265. HELIOTRYPHA PARZUDAKI . . . . Vol. IV. Pl. CCXL.
Ornismya Parzudaki, De Longuem. et Parz.
Mellisuga Parzudaki, Gray & Mitch.
Heliangelus parzudaki, Bonap.
Heliotrypha parzudakii, Bonap.
Trochilus exortis, Fras.
Parzudakia dispar, Reichenb.
* *Ramphomicron dispar*, Reich. Troch. Enum. p. 10.
* *Trochilus lasiopygus*, Licht. in Mus. Berlin.
* *Heliotryphon Parzudakii*, Cab. et Hein. Mus. Hein. Theil iii.p.74.
Habitat. New Granada; and Ecuador, where it is rare.

266. HELIOTRYPHA VIOLA, *Gould* . . . Vol. IV. Pl. CCXLI.
Heliangelus viola, Gould.

Parzudakia viola, Reichenb. Aufz. der Col. p. 12.
Ramphomicron viola, Id. Troch. Enum. p. 10.
Heliotryphon viola, Cab. et Hein. Mus. Hein. Theil iii. p. 74.
Habitat. Ecuador.

Genus HELIANGELUS, *Gould*.

("Ηλιος, sol, et άγγελος, angelus.)

Generic characters.

Male.—Bill straight, about the same length as the head, and cylindrical; *wings* somewhat powerful; *tail* rather round in form and of medium size; *feet* moderately strong: *hind toe* and *nail* the same length as the middle toe and nail; *gorget* luminous, bounded below by a crescent of white.

Female.—Destitute of luminous colouring.

This is perhaps a better-defined genus than any other of those into which the Andean groups of Humming-Birds have been divided. Its characteristics are a moderately long bill surmounted by a band of lustrous colour on the forehead, and a deep luminous gorget separated from the general colour of the body by a semicircular band of white. Like the *Helianthea* and *Heliotryphæ* the species of this form range along the Andes on both sides of the equator.

Dr. Reichenbach, in my opinion, went far out of his way when he separated these birds into three genera—*Trochilus, Anactoria,* and *Diotima.* Had he carefully studied the group from actual specimens, he would have seen that this was unnecessary.

267. HELIANGELUS CLARISSE . . . . Vol. IV. Pl. CCXLII.

Ornismia Clarisse, De Longuem.
Mellisuga Clarissa, Gray & Mitch.
Heliangelus Clarisse and Clarissa, Bonap.
Anactoria Clarissa, Reichenb.
Trochilus Clarissa, Reichenb. Troch. Enum. p. 10.
Heliangelus Clarissæ, Cab. et Hein. Mus. Hein. Theil iii. p. 75.
Anactoria Libussa, Reichenb. Aufz. der Col. p. 12; Id. Troch. Enum. p. 10.

Habitat. The high lands of New Granada. Plentiful in collections from Bogota.

268. HELIANGELUS STROPHIANUS, *Gould* Vol. IV. Pl. CCXLIII.

Trochilus strophianus, Gould.
Mellisuga strophiana, Gray & Mitch.
Heliangelus strophianus, Bonap.
Anactoria Strophiana, Reichenb.
Trochilus Strophiana, Reichenb. Troch. Enum. p. 10.

Habitat. Ecuador.

269. HELIANGELUS SPENCEI . . . . Vol. IV. Pl. CCXLIV.

Trochilus Spencei, Bourc.

Mellisuga Spencei, Gray & Mitch.
Heliangelus Spencei, Bonap.
Diotima Spencei, Reichenb.
Trochilus Spencei, Reichenb. Troch. Enum. p. 10.

Habitat. The ranges of Sierra Nevada de Merida in New Granada.

270. HELIANGELUS AMETHYSTICOLLIS . Vol. IV. Pl. CCXLV.

Orthorhynchus amethysticollis, D'Orb. et Lafres.
Mellisuga amethysticollis, Gray & Mitch.
Trochilus amethysticollis, Tschudi.
Lampornis amethysticollis, Tschudi.
Heliangelus amethysticollis, Bonap., Cabanis.
Anactoria amethysticollis, Reichenb.
Trochilus amethysticollis, Reichenb. Troch. Enum. p. 10.

Habitat. Peru.

271. HELIANGELUS MAVORS, *Gould* . . Vol. IV. Pl. CCXLVI.

Mellisuga Mavors, Gray & Mitch.
Heliangelus mavors, Bonap.
Trochilus Mavors, Reichenb.
Trochilus Mavors, Reichenb. Troch. Enum. p. 10.

Habitat. The Paramos of Portachuela and Zumbador in New Granada.

That almost *terra incognita*, so far at least as its zoological productions are concerned, the Andes of La Paz, has given us, through the researches of M. Warszewicz, one of the most distinct as well as one of the most beautiful forms yet discovered among the Trochilidæ. This remarkable bird is the type of my genus *Diphlogæna*, to which I have since added a second species under the name of *D. Aurora*, with a mark of reservation in case it may prove to be the female of *D. Iris*; for the present, however, I regard it as distinct.

Genus DIPHLOGÆNA, *Gould.*

(δι-, duplex, et φλόγαιρος, flammeus.)

Generic Characters.

Male.—Bill straight and longer than the head; *wing* very long and pointed; *tail* lengthened and deeply forked; *tarsi* short and partially clothed; *feet* small; *hind toe* short; *nails* moderately long and straight; *crown* decorated with several luminous colours.

Female.—Unknown.

272. DIPHLOGÆNA IRIS, *Gould* . . . Vol. IV. Pl. CCXLVII.

Helianthea Iris, *Gould, Bonap.*

Habitat. Andes of Bolivia, between Sorata and Illinani. The locality given me by M. Warszewicz is the province of Huancabamba an Cordilera Solaio, 9000 feet.

273. DIPHLOGÆNA AURORA, *Gould* . Vol. IV. Pl. CCXLVIII.
Hypochrysia Aurora, Reichenb. Aufz. der Col. p. 9.
Coeligena Warszewiczii, Reichenb. Aufz. der Col. p. 23.
Coeligena Warszewiczii, Reichenb. Troch. Enum. p. 4, pl. 690. fig. 4526.

Habitat. Peru; locality the same as *D. Iris*.

Dr. Reichenbach's specific name of *Warszewiczi* must, I believe, give place to that of *Diphlogæna Aurora*, unless his name was proposed prior to the 12th of April, 1853, when I read my paper on this and other new species before the meeting of the Zoological Society of London, as reported in the 'Athenæum' of the 16th of the same month.

The form which appears to me to range next in point of affinity is that of *Clytolæma*. The two members of this genus, unlike their predecessors, which are from the Andes, are natives of the low countries,—one, the *C. rubinea*, being found in Brazil, and, so far as we yet know, confined to the most eastern parts of that country; the other, the beautiful *C. aurescens*, is an inhabitant of the forests of the upper part of the Rivers Madeira and Negro.

Genus CLYTOLÆMA, *Gould*.

(Κλυτὸς, celebris, et λαιμὸς, guttur.)

Generic characters.

Male.—Bill straight and rather longer than the head; *wings* moderately long and pointed; *tail* rather short, and very slightly, *tarsi* partially clothed; *feet* strong; *hind toe* and *nail* shorter than the fore toes and nails; *crown* and *gorget* luminous.

Female.—Destitute of any fine colour.

274. CLYTOLÆMA RUBINEA . . . . Vol. IV. Pl. CCXLIX.
Mellisuga Brasiliensis, gutture rubro, Briss.
Trochilus rubineus, Gmel., Lath., Vieill., Cab.
Ornismya rubinea, Less.
Mellisuga rubinea, Gray & Mitch.
Heliomaster rubineus, Bonap.
Trochilus obscurus, Gmel., Lath. (Cabanis).
———— ruficaudatus, Vieill. Nouv. Dict. d'Hist. Nat. tom. vii. p. 370, tom. xxiii. p. 429.
Cynanthus rubineus, Jard. Nat. Lib. Humming Birds, vol. ii. p. 146.
Heliodoxa rubinea, Reich. Troch. Enum. p. 9, pl. 744. fig. 470–69.
Calothorax rubinea, Burm. Th. Bras. ii. p. 340.

Habitat. The eastern portions of Brazil; common at Rio de Janeiro.

275. CLYTOLÆMA? AURESCENS, *Gould* . . Vol. IV. Pl. CCL.
Trochilus (Lampornis) aurescens, Gould.

Polytmus aurescens, Gray & Mitch.
Lampornis aurescens, Bonap.
Margarochrysis aurescens, Reichenb.
Campylopterus aurescens, Bonap.

Habitat. The forests bordering the Rivers Madeira, Upper Amazon, and Negro.

By some Trochilidists it may be thought that this species should form the type of a distinct genus; but after a careful comparison I believe that I have placed it in its right situation; at the same time I admit that there is some little doubt on the subject.

I next proceed to a group of birds of considerable size, with lengthened straight bills, and the plumage and markings of which render them very conspicuous—the prevailing colours being black and white, relieved by blue and other tints on the crown; they have small and very delicate feet, the colours of which are either rosy or white. I consider them to constitute a very distinct section of the Trochilidæ, and I have much pleasure in adopting for them the generic appellation of *Bourcieria* proposed by the late Prince Charles Bonaparte. All the known species are from the Andes, over which they are spread from the southern part of Peru to the northern part of New Granada.

Genus BOURCIERIA, *Bonap.*

As a typical example of the form, I commence with—

276. BOURCIERIA TORQUATA . . . . . Vol. IV. Pl. CCLI.

Ornismya torquata, Boiss.
Mellisuga torquata, Gray & Mitch.
Bourcieria torquata, Bonap., Reichenb.
**Homophania torquata*, Cab. et Hein. Mus. Hein. Theil iii. p. 79.

Habitat. Columbia. Common in the temperate regions round Bogota.

277. BOURCIERIA FULGIDIGULA, *Gould* . Vol. IV. Pl. CCLII.

Homophania fulgidigula, Cab. et Hein. Mus. Hein. Theil iii. p. 79, note.

Habitat. Ecuador.

278. BOURCIERIA INSECTIVORA.

**Trochilus (Lampornis) insectivorus*, Tschudi, Consp. p. 38, No. 211; Id. Faun. Per. p. 249, t. 28. f. 1.

I observe that M. Cabanis has placed the *T. insectivorus* of Tschudi among the synonyms of *B. torquata*; but, having had Tschudi's type specimen sent to me from Neuchatel, I am not satisfied as to its identity with that species. The specimen referred to seems to me to be the young of some bird of which we have not yet seen the adult. I therefore retain the name in my list; but of course

do not figure it. It appears to me to offer an alliance to the
B. Conradi.

Habitat. Peru.

279. BOURCIERIA CONRADI . . . . . Vol. IV. Pl. CCLIII.

Trochilus Conradii, Bourc.
Mellisuga Conradii, Gray & Mitch.
Conradinia Conradi, Reichenb.
Bourcieria Conradi, Bonap., Reichenb.
Helianthea Conradi, Cab. et Hein. Mus. Hein. Theil iii. p. 80, note.

Habitat. Pamplona in New Granada.

280. BOURCIERIA INCA, *Gould* . . . . Vol. IV. Pl. CCLIV.

Bourcieria Inca, Gould, Bonap., Reichenb.
Homophania Inca, Cab. et Hein. Mus. Hein. Theil iii. p. 79.

Habitat. Province of Coroico in Bolivia; 6000 or 8000 feet (*Warszewicz*).

Genus LAMPROPYGIA, *Reichenb*.

The members of this genus (all figured in the work under the generic appellation of *Cœligena*) bear a general resemblance to the last as regards their size and the lengthened and straight form of their bills; but their style of colouring is very different, and, however much some naturalists may dissent from the idea of colour being regarded as a generic character, I do think that it is of no little importance in this group of birds; for I find that every distinct section or genus is distinguished by some peculiar style of plumage and colouring common to all the species of which it is composed, and not found in the others. Thus the members of the present genus all bear a plumage of a rather dull or sombre character with the exception of the lower part of the back, where it is luminous; but, as is the case with the *Aglæactines*, this luminous colouring is only to be seen when viewed from behind. All the known species are found among the Andes, both on the northern and southern sides of the equator.

281. LAMPROPYGIA CŒLIGENA.

Cœligena typica . . . . Vol. IV. Pl. CCLV.

Ornismya cœligena, Less.
Mellisuga cœligena, Gray & Mitch.
Cœligena typica, Bonap.
Lampornis cœligena, Jard. Nat. Lib. Humming-Birds, vol. ii. p. 156.
Cœligena typica, Reichenb. Troch. Enum. p. 3, pl. 686. fig. 4515.
Lampropygia cœligena, Cab. et Hein. Mus. Hein. Theil iii. p. 78.

Habitat. New Granada.

In my account of this species, which is common in the neighbourhood of Bogota, I stated that the Bolivian birds which appeared to be identical with it are much larger in size and darker in colour, and that I thought it probable that they would prove to be distinct and undescribed; I still entertain the same opinion. I therefore take this opportunity of assigning to this southern representative a specific appellation, but do not consider it necessary to give a figure of it.

282. LAMPROPYGIA BOLIVIANA, *Gould*.

This bird bears a general resemblance to the *L. cæligena*, but differs in being of a much darker colour on the head and neck, and in having the tail dark olive-brown washed with bronze in lieu of light bronzy-brown; the lower part of the back also is more richly coloured, the crescentic markings of green showing still greater lustre when viewed from behind.

Total length 5¼ inches; bill 1⅜; wing 3⅜; tail 2½; tarsi ¼.

Habitat. Bolivia.

283. LAMPROPYGIA PURPUREA.

Cœligena purpurea, *Gould* . . . . . Vol. IV. Pl. CCLVI.
*Cæligena ———, Bonap. Consp. Gen. Av. tom. i. p. 73, *Cæligena*, sp. 2.
Cæligena purpurea, Reich. Troch. Enum. p. 3, pl. 753. figs. 4727, 4728.
Lampropygia purpurea, Cab. et Hein. Mus. Hein. Theil iii. p. 71, note.

Habitat. Popayan.

284. LAMPROPYGIA PRUNELLEI.

Cœligena Prunelli . . . . Vol. IV. Pl. CCLVII.
Trochilus Prunellei, Bourc.
Mellisuga Prunellei, Gray & Mitch.
Homophania Prunellii, Reichenb.
Bourcieria prunelli, Bonap.
Bourcieria Prunelli, Reichenb. Troch. Enum. p. 7, pl. 750. figs. 4721, 4722.
Homophania Prunelli, Cab. et Hein. Mus. Hein. Theil iii. p. 79.

Habitat. Andes of New Granada. Common in collections from Bogota.

285. LAMPROPYGIA WILSONI.

Cœligena Wilsoni . . . . Vol. IV. Pl. CCLVIII.
Trochilus Wilsoni, Delatt. et Bourc.
Mellisuga Wilsoni, Gray & Mitch.
Bourcieria Wilsoni, Bonap., Reichenb.
Lampropygia Wilsoni, Reichenb., Cabanis.

Habitat. Ecuador.

"There must be some error when M. Bourcier states that he killed this species at an elevation of 10,000 feet, and in Nono, which lies at about 9000 feet. The bird belongs strictly to the warmer countries, such as Nanegal, which is only about 4000 feet in altitude."— *Jameson and Fraser* in *Ibis*, vol. i. p. 400.

A group rather than a genus next claims our attention; for two or three very well-marked divisions occur among the birds I have figured under the generic name of *Heliomaster*. Unlike the last, which are confined within certain limits, these birds are widely spread, some of them over Mexico and Central America, and others over Venezuela, and even further south than the latitude of Rio de Janeiro in Brazil.

The members of this section of the Trochilidæ are of rather large size, have long straight bills, lengthened wings, and a structure admirably adapted for aërial progression. The males are mostly clothed with fine colours on the crown and throat.

The species of the

Genus HELIOMASTER, Bonap.,

as now restricted, are at least five or six in number, and four of them are inhabitants of Central America or countries north of the Isthmus of Panama. Their short, nearly square tails, the outer feathers of which, together with their under tail-coverts, are spotted with white, render them very conspicuous.

286. HELIOMASTER LONGIROSTRIS . . . Vol. IV. Pl. CCLIX.

Trochilus longirostris, Vieill.
———— *superbus*, Shaw, Lath., Temm., Jard.
Ornismya superba, Less.
Long-billed Humming-Bird, Lath.
Mellisuga longirostris, Gray & Mitch.
Heliomaster longirostris, Bonap., Reich., Cabanis.
**Selasphorus longirostris*, Reichenb., Troch. Enum. p. 11.

Habitat. Trinidad.

287. HELIOMASTER STUARTÆ, Lawr.

**Heliomaster longirostris*, Sclat. in Proc. of Zool. Soc. part xxv. p. 16.
*———— *Stuartæ*, Lawr. Ann. Lyc. Nat. Hist. in New York, April 9, 1860.

Habitat. New Granada; the neighbourhood of Bogota.

Mr. G. N. Lawrence, of New York, considers the bird from Bogota to be distinct from the *Heliomaster longirostris* of Trinidad, and has assigned to it the distinctive appellation of *Stuartæ*, in honour of a most estimable lady, the wife of R. L. Stuart, Esq., of New York.

For my own part, I have always regarded the Bogota and Trinidad birds as one and the same; but Mr. Lawrence has ever maintained that they are distinct; and on the day when these remarks were written I received from him a copy of the first part of his paper entitled "Catalogue of a Collection of Birds made in

New Granada, by James McLennan, Esq., of New York," in which, referring to his *Heliomaster Stuartæ*, he says:—"Since describing this species, I have had an opportunity of examining seven other specimens from Bogota. I find the bills of these to be quite as long as those of *H. longirostris*; but they are much stouter, and the base of the bill is very broad and bare of feathers, whereas in *longirostris* the bill is comparatively narrow at the base, and the feathers extend quite forward on the bill. These differences were constant in an equal number of each species."

Whether the birds are really distinct or mere local varieties, time and the acquisition of a larger number of specimens must determine. I have in my own collection two specimens of another bird of this form, which bear a very general resemblance both in size and markings; but the crown, instead of being bluish green, is positive blue. So decided is this colour, that I have no hesitation in saying that, if so slight a difference is allowed to separate the Bogotan and Trinidadian birds, these also must be regarded as belonging to a distinct species, and the term *Sclateri*, which has been proposed by Dr. Cabanis, be used for them. I have two very fine males of this bird in fully adult plumage, killed by M. Warszewicz in Costa Rica; but in what particular locality, is unknown to me. Besides the Costa-Rican bird, I have another, still more different, from Southern Mexico, a most charming specimen, killed by M. Montes de Oca. This beautiful bird also bears a general resemblance in colour and markings to those immediately preceding, but is distinguished from all of them by its delicate light-green metallic crown. For this new species I propose the name of *pallidiceps*.

If this little section be found to be composed of four distinct birds, the species will stand thus:—*H. longirostris* of Trinidad and the adjacent portions of the continent; *H. Stuartæ* of Bogota; *H. Sclateri* of Costa Rica; and *H. pallidiceps* of Guatemala and Mexico.

288. HELIOMASTER SCLATERI, *Cabanis*.

Heliomaster Sclateri, Cab. et Hein. Mus. Hein. Theil iii. p. 54, March 30, 1860.

Habitat. Costa Rica.

289. HELIOMASTER PALLIDICEPS, *Gould*.

Crown of the head shining pale green, much paler than in *H. longirostris*; chin black; gorget purplish red, separated from the ear-coverts by a distinct mark of white; upper surface bronzy green; two centre tail-feathers wholly green, the next on each side green tipped with black; the three outer feathers green at the base, then black, and a spot of pure white at the tip, the white spot becoming less from the outer one, until on the third it is a mere speck; wings purplish brown; chest and centre of the abdomen grey; flanks bronzy green; under tail-coverts pale green, tipped with white.

Total length $3\frac{3}{4}$ inches; bill $1\frac{5}{8}$; wing $2\frac{3}{4}$; tail $1\frac{1}{4}$; tarsi $\frac{1}{4}$.

Habitat. Mexico and Guatemala.

"The white sides and the white spot on the back show very conspicuously as this bird rests on its perch."—*Salvin* in *Ibis*, vol. ii. p. 264.

290. HELIOMASTER CONSTANTI . . . Vol. IV. Pl. CCLX.
Ornismya Constanti, Delatt.
Habitat. Guatemala, and Costa Rica.

291. HELIOMASTER LEOCADIÆ.
Heliomaster pinicola, *Gould* . . . . Vol. IV. Pl. CCLXI.
Trochilus Leocadiæ, Bourc. Ann. des Sci. Nat. de Lyon, tom. iv. 1852.
Habitat. Mexico.

Genus LEPIDOLARYNX, *Reich.*

This form, of which the single species known has received the above generic appellation, differs in many particulars from the preceding: the bill is less elongated and not so straight, while the tail is decidedly forked; independently of which, the gular mark is very different, the entire throat being luminous, while in all the species of *Heliomaster* the chin is black.

292. LEPIDOLARYNX MESOLEUCUS.
Heliomaster mesoleucus . . . . . Vol. IV. Pl. CCLXII.
Trochilus mesoleucus, Temm.
————— longirostris, Natt.
————— squamosus, Temm.
————— mystacinus, Vieill.
Ornismya Temminckii, Less.
————— mesoleuca, Less.
Mellisuga mesoleuca, Steph.
————— squamosa, Steph.
————— melanoleuca, Gray & Mitch.
Heliomaster mesoleucus, Bonap.
*Lepidolarynx mesoleucus, Reichenb. Aufz. der Col. p. 13.
*Selasphorus (*Lepidolarynx*) mesoleucus, Reichenb. Troch. Enum. p. 11.
*Ornithomyia mesoleuca, Bonap. in Rev. et Mag. de Zool. 1854, p. 251.
*Heliomaster squamosus, Cab. et Hein. Mus. Hein. Theil iii. p. 53.
*Calothorax mesoleucus, Burm. Th. Bras. ii. p. 339, 1.
*Trochilus mystacinus, Vieill. MSS.
Habitat. Brazil.

Genus CALLIPERIDIA, *Reich.*

The *Calliperidia Angelæ* offers a still further departure from the true *Heliomasters;* for it has even a shorter bill than the last, while its tail is much more deeply forked. It is by far the finest species yet discovered. Its entire body is clothed in glittering colours, and the bird itself must be seen and examined to obtain an idea of its beauty. The female, on the other hand, has the under surface of the body smoky grey, differing in this respect from all the others.

293. CALLIPERIDIA ANGELÆ.

Heliomaster Angelæ . . . . . . Vol. IV. Pl. CCLXIII.
Ornismya Angelæ, Less.
Heliomaster angelæ, Bonap.
Calliperidia Angelae, Reichenb. Aufz. der Col. p. 12.
Calliphlox Angelae, Reichenb. Troch. Enum. p. 10.
Ornithomyia angela, Bonap. Rev. et Mag. de Zool. 1854, p. 251.

Habitat. Buenos Ayres and Tucuman.

A single specimen of a Humming-Bird, somewhat allied to the last genus, was killed by M. Warszewicz on the sides of the Volcano of Chiriqui. It possesses so many remarkable characters that I was obliged, without a moment's hesitation, to assign to it a new generic appellation, that of *Oreopyra*. Unfortunately the specimen was so much injured by shot that I had much difficulty in giving a correct delineation of it.

Genus OREOPYRA, *Gould*.
("Opos, mons, et πῦρ, ignis.").

Generic characters.

Male.—Bill longer than the head, straight, or very slightly arched; *wings* long and rigid; *tail* moderately long and forked; *tarsi* clothed; *feet* rather small; *gorget* snow-white.

294. OREOPYRA LEUCASPIS, *Gould* . . Vol. IV. Pl. CCLXIV.

Habitat. Volcano of Chiriqui, 9000 to 10,000 feet (*Warsewics*.)

So different are the three birds found on the island of Juan Fernandez, that it would not involve a great stretch of impropriety to assign to each of them a separate generic appellation; I shall, however, retain them all under the name of *Eustephanus*. On an examination of the plates of the three species it will be seen how remarkably they differ in size, colour, and markings. I consider it a very singular fact connected with the family of Humming-Birds, that three species should be found on an island so distant from the mainland, and that two of them should be confined to this isolated spot, surrounded as it is by the wide waters of the Pacific Ocean.

Genus EUSTEPHANUS, *Reichenb.*

295. EUSTEPHANUS GALERITUS . . . Vol. IV. Pl. CCLXV.
Trochilus galeritus, Mol., Lath., Buff., Sonn., Vieill.
Mellisuga Kingii, Vig.
Ornismya sephanoïdes, Less. et Garn.
Trochilus sephanoïdes, Jard.

142

Trochilus forficatus, Gould.
———— *flammifrons*, Lyell.
Mellisuga galerita, Gray & Mitch.
Sephanoides galerita, Bonap.
*Sephanoides Kingi, Gray, List of Gen. of Birds, p. 19.
*Eustephanus galeritus, Reichenb. Aufz. der Col. p. 14; Id. Troch.
  Enum. p. 11; Cab. et Hein. Mus. Hein. Theil iii. p. 76.
Habitat. Chili and Juan Fernandez.

"Found about Valparaiso in abundance in the months of August, September, and October."—*Bridges* in *Proc. of Zool. Soc.* part. xi. p. 115.

296. EUSTEPHANUS STOKESI . . . . Vol. IV. Pl. CCLXVI.

Trochilus Stokesi, King, Less., Jard.
Mellisuga Stokesi, Gray & Mitch.
Sephanoides stokesi, Bonap.
*Thaumaste Stokesii, Reichenb. Aufz. der Col. p. 14; Id. Troch. Enum. p. 12.
*Eustephanus Stokesi, Cab. et Hein. Mus. Hein. Theil iii. p. 75.
Habitat. Juan Fernandez.

297. EUSTEPHANUS FERNANDENSIS . Vol. IV. Pl. CCLXVII.

Trochilus Fernandensis, King.
Ornismya cinnamomea, Gerv.
———— *Robinson*, Less.
Mellisuga Fernandensis, Gray & Mitch.
Sephanoides fernandensis, Bonap.
*Eustephanus Fernandensis, Reichenb. Aufz. der Col. p. 14; Id. Troch. Enum. p. 11; Cab. et Hein. Mus. Hein. Theil iii. p. 76.
Habitat. Juan Fernandez.

The

Genus PHÆOLÆMA, Reichenb.,

is composed of two Andean species, distinguished by their sombre colouring; for although both have a luminous gular patch, and one of them a glittering mark on the centre of the crown, the brilliancy of these markings is not so great as usual, and their tails are coloured unlike those of any other group.

298. PHÆOLÆMA RUBINOÏDES . . . Vol. IV. Pl. CCLXVIII.

Trochilus rubinoïdes, Bourc. et Muls.
Mellisuga rubinoïdes, Bonap.
Heliomaster rubinoïdes, Bonap.
Clytolæma rubinoïdes, Bonap.
Phaiolaima rubinoïdes, Reichenb.
*Heliodoxa rubinoïdes, Reichenb. Troch. Enum. p. 6, pl. 743. figs. 4704-5.
Habitat. New Granada. Frequently sent to Europe from Bogota.

299. PHÆOLÆMA ÆQUATORIALIS, *Gould* Vol. IV. Pl. CCLXIX.

Phæolæma rubinoides, Sclat.
———— *æquatorialis*, Gould, Sclat.

Habitat. Ecuador.

Genus ERIOCNEMIS, *Reichenb*.

The conspicuous tufts of feathers with which the legs of the *Eriocnemides* are clothed is a feature both novel and peculiar; and as it is not to be found in any other group of birds, they are thereby rendered especially singular. In some these powder-puff-like decorations are white, in others brown and white, and in one jet black. All the species are confined to that portion of the Andes which is bounded on the north by New Granada, and on the south by Bolivia.

It is not to be supposed that the minor distinctive characters which exist among the many species of this group should have passed unnoticed by ornithologists; on the contrary, they have attracted the notice of more than one writer, and the birds which were all formerly included in the genus *Eriopus* or *Eriocnemis* have received the subgeneric titles of *Engyete*, *Threptria*, *Phæmonoë*, *Aline*, *Luciania*, *Mosqueria*, *Derbyomia*, &c.,—a tolerable division for the genus first established by me in 1847, under the name of *Eriopus*. I shall now give my own views on the subject, and point out those which I consider to be natural divisions. The first, then, is the well-known *E. cupreiventris*, with which I associate the *E. Isaacsoni*, the *E. Luciani* and the *E. Mosquera*. In all these the sexes are alike in colour. The next division comprises *E. vestita* and *E. nigrivestis*, as they both have a brilliant patch of feathers on the throat and the lower part of the back and the upper tail-coverts, exceedingly luminous; and their females are somewhat different and less brilliant in colour. The *E. Godini* and *E. D'Orbignyi* form another little section; but we really know so little respecting these species, that nothing can be said with certainty as to their females. The black-puffed *E. Derbianus* stands alone, and a rare and very beautiful bird it is. The *E. Alinæ* is distinguished from all the rest by the glittering green of its face and under surface; it is by far the smallest species of the genus, while it has the largest puffs; and the female, although bearing a general resemblance to the male, is far less brilliant. The members of the next section are very sombre in their colouring, as will be seen on reference to the plates on which they are represented: they are *E. squamata*, *E. lugens*, and *E. Aureliæ*. Ornithologists may please themselves about adopting generic terms for these minute divisions; but, for myself, I have kept them all under that of *Eriocnemis*, and still feel inclined to do so. They all possess the important character of the puff leg, and they are remarkably alike as to the amount of this peculiar ornamentation.

300. ERIOCNEMIS CUPREIVENTRIS
Vol. IV. Pls. CCLXX., CCLXXI.

Trochilus cupreiventris, Fras.

Ornismya vestita ♀, Longuem.
—— *maniculata*, Less.?
Hylocharis cupreocentris, Gray & Mitch.
Eriopus cupreiventris, Bonap.
Eriopus simplex, Gould.
Eriocnemis simplex, Gould, Bonap. Reichenb.
* *Phæmonoë cupriventris*, Reichenb. Aufz. der Col. p. 9.
* *Eriocnemis cupriventris*, Reichenb. Troch. Enum. p. 6, pl. 729, figs. 4668–69.
* *Eriocnemis cupreiventris*, Cab. et Hein. Mus. Hein. Theil iii. p. 73.

Habitat. The Andes in New Granada.

I now believe that the bird I have called *Eriocnemis simplex* is merely a dark variety of the *E. cupreiventris*. Such varieties do now and then occur with other species of the family; the cause I cannot attempt to explain.

301. ERIOCNEMIS ISAACSONI . . . . Vol. IV, Pl. CCLXXII.

Ornysmia Isaacsoni, Parz.
Hylocharis Isaacsoni, Gray & Mitch.
Eriocnemys isaacsoni, Bonap.
Phæmonoë Isaacsoni, Reichenb.
* *Eriocnemis Isaacsoni*, Reichenb. Troch. Enum. p. 6, pl. 761. fig. 4700.

Habitat. New Granada.

I have never seen any other than the type specimen of this species, which is now in the Derby Museum at Liverpool.

302. ERIOCNEMIS LUCIANI . . . . Vol. IV. Pl. CCLXXIII.

Trochilus Luciani, Boure.
Hylocharis Luciani, Gray & Mitch.
Eriopus luciani, Bonap.
T. (Eriopus) Luciani, Jard.
* *Phæmonoë Luciani*, Reichenb. Aufz. der Col. p. 9.
* *Eriocnemis Luciani*, Reichenb. Troch. Enum. p. 6, pl. 730. figs. 4671–72.

Habitat. Ecuador; western side of Pichincha at an elevation of 10,000 to 12,000 feet (*Jameson*).

303. ERIOCNEMIS MOSQUERA . . . . Vol. IV. Pl. CCLXXIV.

Trochilus Mosquera, Bourc. et Delatt.
Hylocharis mosquera, Gray & Mitch.
Eriopus mosquera, Bonap.
* *Threptria Mosquera*, Reichenb. Aufz. der Col. p. 9.
* *Eriocnemis Mosquera*, Reichenb. Troch. Enum. p. 6, pl. 728. figs. 4664–65.

Habitat. The neighbourhood of Pasto in New Granada (*Delattre*).

304. ERIOCNEMIS VESTITA . . . . . Vol. IV. Pl. CCLXXV.
Ornismya vestita, Longuem., Delatt. et Bourc.
Trochilus uropygialis, Fras.
Ornismya glomata, Less.
Hylocharis vestita, Gray & Mitch.
Eriopus vestita, Bonap.
Eriocnemis vestita, Reichenb.
Eriocnemys vestitus, Bonap.
**Eriocnemis vestita*, Cab. et Hein. Mus. Hein. Theil III. p. 73.
Habitat. The Andes of New Granada. Commonly sent from Bogota.

305. ERIOCNEMIS NIGRIVESTIS . . . Vol. IV. Pl. CCLXXVI.
Trochilus nigrivestis, Bourc.
Eriocnemys nigrivestis, Bonap.
Eriocnemis nigrivestis, Reichenb.
T. vestinigra, Verr. MSS.?
Habitat. Ecuador, environs of Tumbaro (*Bourcier*).

306. ERIOCNEMIS GODINI . . . . Vol. IV. Pl. CCLXXVII.
Trochilus Godini, Bourc.
Eriocnemys godini, Bonap.
Eriocnemis Godini, Reichenb.
Habitat. Ecuador.

307. ERIOCNEMIS D'ORBIGNYI . . Vol. IV. Pl. CCLXXVIII.
Trochilus D'Orbignyi, Bourc.
Phæmonoë D'Orbignyi, Reichenb.
Eriocnemis D'Orbignyi, Reichenb.
Eriocnemys orbignyi, Bonap.
Habitat. Peru or Bolivia.

308. ERIOCNEMIS DERBIANA . . . . Vol. IV. Pl. CCLXXIX.
Trochilus Derbyi, Delatt. et Bourc.
Eriopus Derbyi, Gould.
——— *derbyi*, Bonap.
Eriocnemys derbyanus, Bonap.
Treptria Derbyi, Reichenb.
**Eriocnemis Derbyi*, Reichenb. Troch. Enum. p. 6, pl. 728. figs. 4666-67 and pl. 741. figs. 4698-99.
Habitat. Volcano of Puracé in New Granada (*Delattre*).

309. ERIOCNEMIS ALINÆ . . . . . Vol IV. Pl. CCLXXX.
Ornismyia Alinæ, Bourc.
Ornismya Alinæ, Bourc.
Hylocharis Alinæ, Gray & Mitch.
Eriopus alinæ, Bonap.
Engyets Alinæ, Reichenb.
Eriocnemys alinæ, Bonap.

Trochilus dasypus, Licht. in Mus. of Berlin.
Eriocnemis Alinae, Cab. et Hein. Mus. Hein. Theil iii. p. 79.
Habitat. The hilly parts of New Granada.

310. ERIOCNEMIS SQUAMATA, *Gould* . Vol. IV. Pl. CCLXXXI.
Habitat. Ecuador.

311. ERIOCNEMIS LUGENS, *Gould* . . Vol. IV. Pl. CCLXXXII.
Eriopus lugens, Gould.
Eriocnemys lugens, Bonap.
Threptria lugens, Reichenb. Aufz. der Col. p. 9.
Eriocnemis lugens, Reichenb. Troch. Enum. p. 6, pl. 740. figs. 4695–96.
Habitat. Ecuador; western side of Pichincha (*Jameson*).

It is just possible that this may prove to be the female of *E. squamata*; for I have received many specimens from Professor Jameson with wholly white puffs, which is the characteristic of the *E. lugens*; while from another locality one has been sent with partly white and partly red puffs: independently of the difference in the colouring of the puffs, the latter birds are larger than the former.

312. ERIOCNEMIS AURELIÆ . . . Vol. IV. Pl. CCLXXXIII.
Trochilus Aureliæ, Boure.
Hylocharis Aureliæ, Gray & Mitch.
Eriopus aureliæ, Bonap.
Eriocnemys aureliæ, Bonap.
Eriocnemis Aureliæ, Reichenb.
Habitat. New Granada and Ecuador.

Specimens from the Napo differ considerably from those received from Bogota,—a deep coppery hue pervading both the upper and under surface, whereas those parts are green in the Bogotan birds. I have seen specimens which I consider may be females or young of this species with wholly white puffs.

Proceeding from Mexico, southwards, through the high lands of the temperate regions of Guatemala, Costa Rica, and Veragua, we there find several species of the well-defined genus *Cyanomyia* which do not pass the Isthmus of Panama, while others occur in New Granada, Ecuador, and Peru. I have not yet seen any species of this form from Brazil or from any of the eastern portions of the South American continent. They are all very lovely birds, the colours with which they are adorned being blue, glittering green, and white, to which the red bills of one or two of them offer a pleasing contrast. The females, although generally resembling the males, are inferior to them in size and colouring. With these birds I commence the fifth volume.

Genus CYANOMYIA, *Bonap.*

313. CYANOMYIA QUADRICOLOR . . Vol. V. Pl. CCLXXXIV.
Trochilus quadricolor, Vieill.
Polytmus quadricolor, Gray & Mitch.
Cyanomyia quadricolor, Bonap.
Uranomitra quadricolor, Reichenb.
* *Ornismya cyanocephala*, Less. Supp. des Ois.-mou. p. 132, pl. 17.
* *Trochilus verticalis*, Licht. Preis-Verz. Mexican. Thier. gen. v.
   Deppe & Schiede (Sept. 1830), Nos. 27, 28.
* *Cyanomyia verticalis*, Bonap. Rev. et Mag. de Zool. 1854, p. 254.
* *Agyrtria quadricolor*, Reichenb. Troch. Enum. p. 7, pl. 761. figs.
   4758–59.
* *Uranomitra quadricolor*, Cab. et Hein. Mus. Hein. Theil iii. p. 41.

Habitat. Northern Mexico.

314. CYANOMYIA VIOLICEPS, *Gould* . Vol. V. Pl. CCLXXXV.
* *Uranomitra violiceps*, Cab. et Hein. Mus. Hein. Theil iii. p. 41, note.

Habitat. Western Mexico.

315. CYANOMYIA CYANOCEPHALA . . Vol. V. Pl. CCLXXXVI.
* *Ornismya cyanocephala*, Less. Supp. des Ois.-mou. p. 134, pl. 18.
* *Polytmus verticalis*, Gray & Mitch. Gen. of Birds, vol. i. p. 109.
   Polytmus, sp. 86.
* ? *Uranomitra cyanocephala*, Reichenb. Aufz. der Col. p. 10.
* *Cyanomyia cyanocephala*, Bonap. Rev. et Mag. de Zool. 1854,
   p. 254; Montes de Oca, Proc. Acad. Nat. Sci. Philad. 1860,
   p. 80.
* *Agyrtria Faustine*, Reichenb. Troch. Enum. p. 7, pl. 760. figs.
   4756–57.
* *Agyrtria cyanocephala*, Reichenb. ib. p. 7, pl. 760. figs. 4754–55.
* *Uranomitra Lessoni*, Cab. et Hein. Mus. Hein. Theil iii. p. 41.

Habitat. Southern Mexico.

"This Humming-Bird," says M. Montes de Oca, "is generally known by the name of *Chupa-mirto comun de peche blanco*, or Common White-breasted Myrtle-sucker. It is found very abundantly and at all seasons of the year in the vicinity of Jalapa, Coatepec, Orizaba, and many other places in Mexico, where it remains all the year round, and I have often found its nest in the months of April and May.

"This pretty little bird is very familiar and unsuspicious, will allow of a near approach in the woods, and is a constant visitor of the gardens of the towns and cities. Like the fine *Campylopterus Delattrei*, it frequents the *magapan* flowers, around which it may be seen at all hours of the day.

"The nest of this species, like those of nearly all the Humming-Birds of this part of Mexico (Jalapa), is lined with the *tull* silky floss; and is most ingeniously covered on the outside with moss from the rocks. The eggs are generally two in number, but upon one occasion I found three in a nest."

In my account of this species, I have stated that it was found in Guatemala as well as Mexico; and although this may be the case, I believe the latter country to be its true habitat. Guatemalan differ from Mexican specimens in the colouring of the tail-feathers; in the former they are olive-green, in the latter rich bronzy-green. Under these circumstances I cannot regard them as identical, I must therefore give the Guatemalan bird a distinctive appellation :—

316. CYANOMYIA GUATEMALENSIS, *Gould*.

Cyanomyia cyanocephala, *Salvin* in *Ibis*, vol. ii. p. 39.

Habitat. Guatemala.

"About Dueñas," says Mr. Salvin, "this is an abundant species. It frequents the shrubby forest, feeding principally among the flowers of a tree which abounds there. This tree, which grows to a height of about 20 or 30 feet, bears clusters of white flowers, and has its branches and stem covered with spines, which sting when touched. Its bark also, when bruised, emits a milky fluid, which blisters the skin if any be allowed to remain upon it. The bird, when taking its food from this tree, places itself in front of a bunch of the flowers, and hovers opposite, at a distance of about two or three inches. On perceiving the object of its search, it darts in, and seizing whatever that may be, insect or honey, returns to its position in front of the cluster. So it passes on from blossom to blossom, and in like manner from cluster to cluster, until the whole tree is thoroughly ransacked. Humming-Birds do not remain long on the wing at once, but rest, frequently choosing for that purpose a small dead or leafless twig at the top or just within the branches of the tree. While in this position they take the opportunity of trimming their feathers and cleaning their bill, all the time keeping up an incessant jerking of their wings and tail. When this operation has been performed, they peer about for fresh flowers at which to dart. The cry of the present species is somewhat represented by the word '*chirik*' uttered frequently and with great rapidity. This cry seems common to all the family; and it is only from an intimate acquaintance that one can trace a difference between the species. When they are flying from one place to another, or pursuing each other, this cry is especially used, and in the latter case it is uttered with great vehemence. The humming sound from which these birds take their trivial name is something like that produced by a large beetle; but very little practice will soon so accustom the ear that it seldom mistakes the unseen presence of a Humming-Bird for anything else."—*Ibis*, vol. i. p. 127.

"Last year, in a cypress tree near the house at Dueñas, a pair of these birds built their nest. This year I found a branch of the same tree similarly tenanted, the new nest being only a few yards from the site of the old one. To obtain it, I was obliged to cut away the branch; and though, in falling, the nest was quite thrown on its side, the eggs, much to my surprise, did not fall out; this I afterwards found was owing to the lip of the nest turning inwards. Another pair commenced building near the house; and the male bird frequently came while I was preparing skins in the corridor, and took pieces of cotton almost from my hand. In the afternoon of August 14 my friend Mr. Wyld observing it making a descent upon some small object in his room, shut the window and called me. The intruder, who was wearied from fluttering against the window, suffered itself to be caught. In a very few moments its agitation ceased, and it seemed to be taking advantage of its comfortable place in my hand to rest from its fatigues, making no attempt to escape. Before letting it go I procured a piece of sugar, and, after dipping it in water, put it to the tip of its bill. Almost immediately its long tongue was employed in sucking up the liquid. On liberating it, it flew to a tree close at hand, and seemed to take no further notice of its late captivity."—*Salvin* in *Ibis*, vol. ii. p. 39.

317. CYANOMYIA FRANCIÆ. . . . . Vol. V. Pl. CCLXXXVII.

Trochilus Franciæ, Bourc.
Uranomitra Franciæ, Reichenb.
Cyanomyia franciæ, Bonap., Selat.
Polytmus Franciæ, Gray & Mitch.
Agyrtria Franciæ, Reichenb. Troch. Enum. p. 7, pl. 761. figs. 4760–61).
Trochilus hypoleucus, Licht. in Mus. of Berlin.
Uranomitra Fruciæ, Cab. et Hein. Mus. Hein. Theil iii. p. 41.

Habitat. New Granada.

318. CYANOMYIA CYANICOLLIS, *Gould* Vol. V. Pl. CCLXXXVIII.

Trochilus (———?) *cyanocollis*, Gould.
Uranomitra cyanicollis, Reichenb.
Cyanomyia cyanicollis, Bonap.
Agyrtria cyanicollis, Reichenb. Troch. Enum. p. 7.
Uranomitra cyanicollis, Cab. et Hein. Mus. Hein. Theil iii. p. 41, note.

Habitat. Peru.

Nearly allied to *Cyanomyia* is the

Genus HEMISTILBON, *Gould.*

('Ημι-, semi, et στίλβων, micans.)

Generic characters.

Male.—Bill longer than the head and straight; *wings* moderately long and considerably curved; *tail* rather short and truncate; *tarsi* clothed; *feet* rather small; *hind toe* shorter than the middle toe; *nails* short and curved.

Although I have placed this genus next to *Cyanomyia*, I consider that it has some relationship to the *Amazilia*.

319. HEMISTILBON OCAI, *Gould*.

Amazilia Ocai, *Gould* . . . . . Vol. V. Pl. CCLXXXIX.
Pyrrhophaena Ocai, Cab. et Hein. Mus. Hein. Theil. lii. p. 36, note.

Habitat. Mexico.

This beautiful species was discovered by M. Rafael Montes de Oca at Xalapa.

In this genus I provisionally place the *Trochilus Norrisi*, of which a single specimen exists in the Loddigesian collection, but unfortunately is so situated that I could not subject it to so careful an examination as I could have desired.

320. HEMISTILBON NORRISI.

* *Trochilus Norrisii*, Bourc. Proc. of Zool. Soc. part xv. p. 47.
* *Polytmus Norrisi*, Gray and Mitch. Gen. of Birds, vol. i. p. 108; *Polytmus*, sp. 71.
* *Amazilia Norrisii*, Reichenb. Aufz. der Col. p. 10; Id. Troch. Enum. p. 8.
* *Amazilius Norrisi*, Bonap. Rev. et Mag. de Zool. 1854, p. 254.
* *Pyrrhophaena Norrisi*, Cab. et Hein. Mus. Hein. Theil iii. p. 36, note.

Habitat. Bolanos in Central Mexico.

At present only one species is known of the

Genus LEUCIPPUS, *Bonap.*;

but I am inclined to believe that other birds of this form exist; indeed I have all but positive evidence that such is the case, in a specimen killed by M. Warszewicz in Peru, which for the present I decline describing, as it may possibly be only a female of some unknown species, the male of which will be differently coloured.

321. LEUCIPPUS CHIONOGASTER . . . . . Vol. V. Pl. CCXC.

Trochilus leucogaster, Tschudi.
Lampornis chionogaster, Tschudi.
Polytmus chionogaster, Gray & Mitch.
Leucippus turneri, Bonap.
Thaumatias leucogaster, Bonap.
Trochilus Turneri, Bourc.
——— (——?) *hypoleucus*, Gould.
Polytmus hypoleucus, Gray & Mitch.
Thaumantias chionogaster, Bonap.
Leucippus Turneri, Reichenb.
* *Leucippus chionogaster*, Cab. et Hein. Mus. Hein. Theil lil. p. 91.

Habitat. Peru and Bolivia.

Genus LEUCOCHLORIS, *Reichenb.*

This is another form of which a single species only has yet been discovered.

322. LEUCOCHLORIS ALBICOLLIS . . . . Vol. V. Pl. CCXCI.

Trochilus albicollis, Vieill., Temm., Less., Jard.
Ornismya albicollis, Less.
Lampornis ulbicollis, Less.
Basilinna albicollis, Less.
Colibri albigularis, Spix.
Polytmus albicollis, Gray & Mitch.
Thaumatias albicollis, Bonap.
Thaumantias albicollis, Bonap.
Leucochloris albicollis, Reichenb.
* *Trochilus vulgaris*, Wied, Beit. iv. p. 72.
* *Leucippus albicollis*, Reichenb. Troch. Enum. p. 8, pl. 782. figs. 4818–19.
* *Agyrtria albicollis*, Cab. et Hein. Mus. Hein. Theil iii. p. 32.
* *Thaumatias albicollis*, Burm. Th. Bras. tom. ii. p. 342.

Habitat. Brazil; and Tucuman, according to Dr. Burmeister.

This is a very pretty species, distinguished by its green and white plumage. The sexes are alike in external appearance.

The genus *Thaumatias*, as proposed by Bonaparte and adopted by me, comprises many species respecting which much confusion exists, especially with regard to the names applied to them by the older authors; some confusion, also, occurs with respect to the species I have called *Thaumatias Linnæi*, which I fear cannot be satisfactorily unravelled. If it should be decided that it is not the bird Bonaparte intended, I trust it will be allowed to stand as the " *Thaumatias Linnæi*, Gould, nec Bonap." The other synonyms may or may not be applicable to it, but they are given on the authority of M. Bourcier of Paris. This is another of those instances which unfortunately occur too frequently for the advantage of science; I mean the impossibility of determining the species intended in the curt descriptions left us by Gmelin and others of *Tobaci* or *Tobagensis*, *Ourissia*, cum multis aliis.

Genus THAUMATIAS, *Bonap.*

The species of this form are natives of Brazil, the banks of the Rivers Amazon and Napo, the Guianas, the Island of Trinidad, Venezuela, New Granada, and Central America. The sexes of each species are alike in colour. If any difference be observable, it is in the outer tail-feathers of the female being faintly tipped with olive-grey.

323. THAUMATIAS CANDIDUS . . . . . Vol. V. Pl. CCXCII.

Trochilus candidus, Bourc. et Muls.
Polytmus candidus, Gray & Mitch.
Thaumatias candidus, Bonap.
* *Agyrtria candida*, Cab. et Hein. Mus. Hein. Theil iii. p. 33, note.

Habitat. Guatemala and Southern Mexico.

Mr. Salvin states that this species is "common on the Atlantic coast-region, about Yzabal, and thence one day's ride into the interior. Very abundant about Coban. Many species of Humming-Birds in Guatemala extend through a great range of temperature, the same species being frequently found both in the coast regions and also in the more elevated districts." Mr. Taylor saw very few examples of this bird in Honduras.

324. THAUMATIAS CHIONOPECTUS, *Gould* Vol. V. Pl. CCXCIII.
  Agyrtria nivcipectus, Cab. et Hein. Mus. Hein. Theil iii. p.39, note.
  Habitat. Trinidad, Cayenne, and Guiana.

325. THAUMATIAS LEUCOGASTER . . . Vol. V. Pl. CCXCIV.
  Trochilus leucogaster, Gmel., Lath., Vieill.
  Mellisuga cayanensis, ventre albo, Briss.
  Agyrtria leucogastra, Reichenb.
  Thaumantias leucogaster, Bonap.
  Ornismya albirostris, Less.
  Cynanthus leucogaster, Jard. Nat. Lib. Humming Birds, vol. ii. p. 149.
  Trochilus mellisugus, Burm. Th. Bras. tom. ii. p. 343.
  Agyrtria leucogastra, Cab. et Hein. Mus. Hein. Theil iii. p. 54.
  Trochilus Cleopatra, Gould, MS.
  Habitat. North-eastern Brazil, Cayenne, and the Guianas.

326. THAUMATIAS VIRIDICEPS, *Gould* . . Vol. V. Pl. CCXCV.
  Habitat. Ecuador.

327. THAUMATIAS MILLERI . . . . . Vol. V. Pl. CCXCVI.
  Trochilus Milleri, Lodd., Bourc.
  Polytmus Milleri, Gray & Mitch.
  Thaumatias Milleri, Bonap.
  Agyrtria Milleri, Reichenb.
  Thaumantias Milleri, Bonap.
  Agyrtria Milleri, Cab. et Hein. in Mus. Hein. Theil iii. p. 33, note.
  Habitat. New Granada. Common in the neighbourhood of Bogota.

328. THAUMATIAS NITIDIFRONS, *Gould* Vol. V. Pl. CCXCVII.
  Habitat. Unknown: supposed to be Venezuela.

329. THAUMATIAS CÆRULEICEPS, *Gould.*
  Thaumatias cæruleiceps, Gould in Proc. Zool. Soc. part xxviii. p. 307.
  Habitat. New Granada. Received from Bogota.

330. THAUMATIAS BREVIROSTRIS . . Vol. V. Pl. CCXCVIII.
  Ornismya brevirostris, Less.
  Basilinna brevirostris, Less.
  Polytmus brevirostris, Gray & Mitch.
  Thaumatias brevirostris, Bonap.
  Thaumantias brevirostris, Bonap.

Agyrtria brevirostris, Reichenb.
* *Trochilus versicolor*, "Licht," Nord. Erm. Reis. Atl. pp. 3, 27, t. 1. f. 1–3.
* *Hylocharis versicolor*, Gray & Mitch. Gen. of Birds, vol. I. p. 115; *Hylocharis*, sp. 45, Bonap. Consp. Gen. Av. tom. I. p. 74; *Hylocharis*, sp. 9.
* *Agyrtria versicolor*, Reichenb. Aufz. der Col. p. 10; Id. Troch. Enum. p. 7, pl. 759. figs. 4750–51.
* *Thaumantias versicolor*, Bonap. Rev. et Mag. de Zool. 1854, p. 255.
* *Agyrtria brevirostris*, Cab. et Hein. Mus. Hein. Theil iii. p. 34.

Habitat. South-eastern Brazil.

331. THAUMATIAS AFFINIS, *Gould* . . Vol. V. Pl. CCXCIX.
* *Agyrtria affinis*, Cab. et Hein. Mus. Hein. Theil iii. p. 33, note.

Habitat. Southern Brazil; the districts of Novo Friburgo and Minas Geraes.

332. THAUMATIAS CHIONURUS, *Gould* . . Vol. V. Pl. CCC
Trochilus (*Thaumatias?*) *chionura*, Gould.
* *Leucippus chionurus*, Reichenb. Aufz. der Col. p. 11; Id. Troch. Enum. p. 8, pl. 780. figs. 4813–15.
* *Thaumantias chionurus*, Bonap. Rev. et Mag. de Zool. 1854, p. 255.
* *Agyrtria chionura*, Cab. et Hein. Mus. Hein. Thell III. p. 32, note.

Habitat. Costa Rica.

333. THAUMATIAS ALBIVENTRIS . . . . Vol. V. Pl. CCCI.
Trochilus tephrocephalus, Vieill.
Ornismya tephrocephalus, Less.
——————— *albiventris*, Less.
Polytmus thaumatias, Gray & Mitch.
Cœligena tephrocephala, Reich.
Agyrtria albiventris, Reich.
Thaumatias albiventris, Bonap.
Thaumantias albiventris, Bonap.
* *Trochilus albiventris*, Jard. Nat. Lib. Humm. Birds, vol. ii. p. 141.
* *Agyrtria albiventris*, Cab. et Hein. Mus. Hein. Theil iii. p. 32.

Habitat. Brazil, the neighbourhood of Sta. Catharina (*Bourcier*).

This, the largest species of the genus, has the centre of the abdomen and the under tail-coverts white.

334. THAUMATIAS LINNÆI, *Bonap.* . . . Vol. V. Pl. CCCII.
Thaumantias linnæi, Bonap.
Trochilus Tobaci, Gmel.
——————— *Tobagensis*, Lath.
——————— *maculatus*, Aud. et Vieill.?
——————— *Tobago*, Shaw?
Ornismya viridissima, Less.

Trochilus viridissimus, Jard. Nat. Lib. Humming Birds, vol ii. p. 149.
Trochilus viridipectus, Sauc. in Mus. Berol. et Heinean.
Saucerottia viridipectus, Reichenb. Aufz. der Col. p. 7.
Agyrtria Thaumantias, Reichenb. Troch. Enum. p. 7, pl. 756. figs. 4738-39.
Chlorestes viridipectus, Relcb. Troch. Enum. p. 4, pl. 702. figs. 4579-75.
Hylocharis lactea, fem., Reich. Troch. Enum. p. 8, pl. 772. fig. 4792.
Coeligena Maugei, fem. (!!!), Reichenb. in Mus. of Berlin (Cabanis).
Agyrtria maculata, Cab. et Hein. Mus. Hein. Theil iii. p. 39.

Habitat. Northern Brazil, Cayenne, Dutch and British Guiana, Trinidad, and Venezuela.

This bird is much smaller than the last, has less white on the abdomen, and the under tail-coverts tinged with grey.

I think it likely that pl. 62 of Lesson's ' Histoire Naturelle des Oiseaux-mouches' has reference to this species, and the text to the *T. albiventris*.

335. THAUMATIAS PLUVIATILIS, *Gould.*

Habitat. Banks of the River Napo.

In size this bird rather exceeds the last, and has a more than proportionately longer bill; the breast is glittering deep or grass green instead of golden green, and there is a narrow irregular streak of greyish white on the abdomen; but the great difference which distinguishes it from the *T. Linnæi* and the rest is the colouring of the under tail-coverts, the centres of which are dark brown margined with greyish white; the tail-feathers are short and of a nearly uniform dull black.

'Total length 4 inches, bill $2\frac{3}{16}$, wing 1, tail $1\frac{1}{4}$.

336. THAUMATIAS APICALIS, *Gould.*

Habitat. New Granada.

About the same size as the last, with an equally lengthened bill; the upper surface golden green; the centre of the abdomen and under tail-coverts pure white; the four outer tail-feathers steel-black margined with pure white at the tip.

Total length $3\frac{1}{4}$ inches, bill 1, wing $2\frac{1}{5}$, tail $1\frac{1}{4}$.

337. THAUMATIAS MACULICAUDUS, *Gould.*

Habitat. British Guiana.

This is a very little species with a long thin bill; its breast is green as in the others; the centre of its abdomen white; the under tail-coverts white, except in the centre, where they are dark brown; but the great difference is in the tail, which is exceedingly pretty, the two centre feathers being bronzy green, except at the extreme tip, which is greenish black, the next on each side bronze for half their length, then black; the three outer ones on each side bronzy green at their base, then broadly zoned with black, next to which they are green, and lastly white.

Total length 3½ inches, bill ⅞, wing 2, tail 1¼.

A specimen of this bird was brought from Guiana by Sir Robert Schomburgk; and in all probability the species is an inhabitant of the interior.

The five preceding species are all very nearly alike, and form a minor section; they are confined to a comparatively limited area. The remaining members of the genus are more widely distributed, one of them inhabiting Guatemala and Costa Rica, and two or three Veragua, while the others frequent Venezuela, Trinidad, the Guianas, and Brazil; but as each of the species has its proper habitat indicated, it will be unnecessary to say more on the subject here. The females are very similar to the males in colour, except in the case of the rare species I have called *T. chionurus*, where the two sexes differ considerably, as may be seen on reference to the plate in which they are represented.

The group of Humming-Birds to which I next direct attention comprise the genera *Amazilia, Pyrrhophæna, Erythronota, Eupherusa, Saucerottia, Sapphironia, Hylocharis,* and others, ranging between the *Amazilia* and the little green species forming the genus *Chlorostilbon*. These birds are the least understood of the Trochilidæ, and are certainly the most difficult and perplexing to discriminate of the entire family. I will, however, do my best to unravel the confusion with which they are surrounded, and to place both the genera and species in as clear a light as my experience enables me. To do this effectively it will be necessary to replace some of the species in the genera from which they have been separated, and to propose a further subdivision of the remainder. In so doing it must not be understood that I am desirous of increasing the number of genera; I merely wish to indicate by a distinctive appellation the sections into which the birds appear to be naturally divided. To particularize the provinces of South America over which the members of the various genera are distributed would be useless; for their dispersion may be said to be general, as they are found from Mexico to Bolivia on the western coast, and from Brazil to Venezuela on the eastern; few of the species, however, go very far either north or south, and still fewer are found in the West India Islands. The members of the genus *Amazilia*, as restricted, are all of somewhat large size, and are easily recognized, each of them having well-marked characters. There is but little difference in the outward appearance of the sexes. The equatorial regions of the Andes are their head-quarters; and it is there that we find the *A. pristina,* the *A. alticola,* the *A. Dumerili,* and the *A. leucophæa.* These four species, I consider, form a very natural section.

Genus AMAZILIA, Reichenb.

338. AMAZILIA PRISTINA . . . . . . Vol. V. Pl. CCCIII.
Orthorhynchus Amazili, Less.

Ornismya Amazili, Less.
Polytmus Amazili, Gray & Mitch.
Trochilus (Lampornis) Amazilia, Tschudi.
Amazilius latirostris, Bonap.
Amazilia latirostris, Reich.
* *Phaëthornis Amazili,* Jard. Nat. Lib. Humming Birds, vol. ii. p. 152.
* *Pyrrhophaena Amazilia,* Cab. et Hein. Mus. Hein. Theil iii. p. 35.
Habitat. The neighbourhood of Lima in Peru.

339. AMAZILIA ALTICOLA, *Gould* . . . . Vol. V. Pl. CCCIV.
Habitat. The high lands of Central Peru; the precise locality uncertain.

340. AMAZILIA DUMERILI . . . . . . Vol. V. Pl. CCCV.
Ornismya Dumerilii, Less.
Trochilus amazicula, Sauc.
Amazilia Amazilicula, Reich.
Polytmus Dumerilii, Gray & Mitch.
Amazilia Dumerilii, Reich.
Amazilius dumerili, Bonap.
* *Pyrrhophaena Dumerili,* Cab. et Hein. Mus. Hein. Theil iii. p. 36, note.

Habitat. Ecuador, on the coast in the neighbourhood of Guayaquil, and on the Isle of Puna. Found also at Babahoyo by Mr. Fraser, who states that the bill is red with a black tip.

341. AMAZILIA LEUCOPHAEA, *Reichenb.* . . Vol. V. Pl. CCCVI.
Amazilia leucophaea, Reichenb.
* *Pyrrhophaena leucophaea,* Cab. et Hein. Mus. Hein. Theil iii. p. 35.

Habitat. Southern Peru. Collected in the vicinity of the Volcano of Arequipa by M. Warszewicz.

I retain Dr. Cabanis's generic term *Pyrrhophaena* for the ten succeeding species:—

342. PYRRHOPHAENA CINNAMOMEA.
Amazilia corallirostris . . . . Vol. V. Pl. CCCVII.
Trochilus corallirostris, Bourc. et Muls.
Polytmus corallirostris, Gray & Mitch.
Amazilius corallirostris, Bonap.
Amazilia corallirostris, Reich.
* *Ornismya cinnamomea,* Less. Rev. Zool. 1842, p. 175.
* *Ornismya rutila,* Delatt. L'Echo du Monde Savant, No. 45 Juin 15, 1843, col. 1069.
* *Pyrrhophaena corallirostris,* Cab. et Hein. Mus. Hein. p. 35, note.
Habitat. Central America.

" This species," says Mr. Salvin, " seems to be an inhabitant of the hot sea-bord only, and does not extend its vertical range to a greater elevation than 2000 feet. In such regions on the Pacific coast it is very abundant, and is, in fact, the commonest of the family—in some parts almost swarming. In every village numbers may be seen flitting about the blossoms of the orange and lime trees. Its horizontal range appears to be extensive, and may be said to include the whole of the southern portion of Guatemala, from the confines of Chiapas to the State of San Salvador, and probably also embraces the Balsam Coast of that republic, as Captain Taylor obtained examples on Tigré Island in the Bay of Fonseca."—*Ibis*, vol. i. p. 130.

"It is common about San Gerónimo; but seems not to be found in the colder and more elevated portions of the republic, neither occurring at Dueñas nor Coban. A nest with two young and the hen bird was brought to me Dec. 6th; the young were half-grown, and would have flown in about ten days. My specimens show that, as far as the feathers are concerned, the sexes are alike. A difference, however, exists in the bill,—that of the male having much more of the brilliant colour from which the species takes its name, in the upper mandible. In the young bird the upper mandible is black."— *Ibis*, vol. ii. pp. 268, 269.

Mr. Taylor, speaking of the birds observed by him in the Republic of Honduras, &c., says, "This Humming-Bird is the only one I observed in any numbers, and that only in certain localities. There were some in Tigré Island, and I saw them here and there on our march across the country. I found them most abundant near Comayagua, 1900 feet above the sea. They were very plentiful on the plain near the town, and not far from the Campo Santo, where the ground was tolerably open and the cactus grew abundantly. There I observed hundreds hovering about the flowers of the cactus."—*Ibis*, vol. ii. p. 115.

343. PYRRHOPHÆNA YUCATANENSIS.

Amazilia Yucatanensis . . . . . . Vol. V. Pl. CCCVIII.

Trochilus Yucatanensis, Cabot.

Habitat. Yucatan.

344. PYRRHOPHÆNA CERVINIVENTRIS, *Gould.*

Amazilia cerviniventris, *Gould* . . . . Vol. V. Pl. CCCIX.

Pyrrhophaena cerviniventris, Cab. et Hein. Mus. Hein. Theil iii. p. 36, note.

Habitat. Mexico. In the neighbourhood of Cordova, according to M. Sallé.

345. PYRRHOPHÆNA CASTANEIVENTRIS.

Amazilia castaneiventris, *Gould* . . . . Vol. V. Pl. CCCX.

Amazilius castaneiventris, Gould.

Habitat. New Granada.

346. PYRRHOPHÆNA RIEFFERI.
Amazilia Riefferi . . . . Vol. V. Pl. CCCXI.
Trochilus Riefferi, Bourc.
Amazilius riefferi, Bonap., Sclat.
Amazilia Riefferi, Reich., Sclat., Salv.
Polytmus Riefferi, Gray & Mitch.
Trochilus Dubusii, Bourc.
Amazilia Dubusii, Reich.
Amazilius dubusi, Bonap., Sclat.
Trochilus fusciccaudatus, Fras.
Hylocharis fusciccaudatus, Gray & Mitch.
* *Ornismya amazili*, Delatt. Echo. du Monde Savant, No. 45, Juin 15, 1843, col. 1069.
* *Trochilus arsinoides*, Sauc. in Mus. of Berlin.
* *Pyrrhophaena Riefferi*, Cab. et Hein. Mus. Hein. Theil iii. p. 36.
* ——————— *Dubusi*, Cab. et Hein. Ib. p. 36.
* ——————— *suavis*, Cab. et Hein. Mus. Hein. Theil iii. p. 36, note.

Habitat. Southern Mexico, Guatemala, and along the Andes to Ecuador.

Nearly thirty specimens are now before me from these various countries, among them M. Bourcier's type specimen of *Dubusi*, also specimens collected by Warszewicz in Costa Rica; and I see nothing to induce a belief that there is any specific difference between those found in Mexico, Guayaquil in Ecuador, or in any of the intermediate countries. I admit that differences occur both in size and in the fringing of the outer tail-feathers: generally speaking, they are darker in the Costa Rican and New Granadian specimens; but I have some quite as bronzy from those countries as the generality of specimens found in Honduras and Guatemala. These latter are the birds to which the term *Dubusi* has been applied.

347. PYRRHOPHÆNA BERYLLINA.
Amazilia beryllina . . . . Vol. V. Pl CCCXII.
Trochilus beryllinus, Licht.
Ornismya Arsinoë, Less.
Cynanthus Arsinoë, Jard.
Polytmus Arsinoë, Gray & Mitch.
Amazilius arsinoë, Bonap.
Amazilia Arsinoë, Reich.
Pyrrhophaena beryllina, Cab. et Hein.

Habitat. Southern Mexico. M. Botta found it at Orizaba, and M. Sallé at Cordova.

348. PYRRHOPHÆNA DEVILLEI.
Amazilia Devillei . . . . Vol. V. Pl. CCCXIII.
Trochilus Devillei, Bourc., Gray & Mitch.

Amazilia Devillei, Reich.
Amazilius devillei, Bonap.
Trochilus Mariæ, Bourc.?
Hylocharis mariæ, Gray & Mitch., Bonap.
Smaragdites Mariae, Reich.
* *Saucerottia maria*, Bonap. Rev. et Mag. de Zool. 1854, p. 255?
Chlorestes Mariae, Reicheub. Troch. Enum. p. 4, pl. 695. fig. 4549.
* *Panychlora Moriae*, Cab. et Hein. Mus. Hein. Theil iii. p. 49, note.
Amazilia Dumerili, Salv. Ibis, vol. ii. p. 270.

Habitat. Guatemala.

Speaking of this species, which by some inadvertence he has called *Dumerili* instead of *Devillei*, Mr. Salvin says, "During the months of July, August, and September, one of its most favourite resorts was the western boundary of the Llaño of Dueñas, which, starting from the village and bounded to the eastward by the river Guacalate, extends, sweeping by the Volcan de Fuego, almost to the Hacienda of Capertillo, its southern extremity. Dispersed all over this plain is found, in groves, patches, and isolated trees, a Tree Convolvulus, bearing a white flower, and attaining an average height of about 25 or 30 feet. During the above months this elegant species might be seen in almost every tree, some feeding among the flowers, some settled quietly on a dead branch, uttering their low, plaintive, hardly to be called musical, yet certainly cheering song, others less peacefully occupied in a war of expulsion, driving out by vehement cries and more effectual blows the tenant of a tree, which in its turn wreaks vengeance on some weaker or unexpectant antagonist."—*Ibis*, vol. ii. p. 270.

349. PYRRHOPHÆNA VIRIDIGASTER.

Amazilis viridigaster . . . . Vol. V. Pl. CCCXIV.
Trochilus viridigaster, Bourc.
Hylocharis viridigaster, Bonap., Gray & Mitch.
Saucerottia viridiventris, Reichenb.
—— *viridigastra*, Bonap., Sclat.
* *Chlorestes viridiventris*, Reichenb. Troch. Enum. p. 4, pl. 699. figs. 4564-65.
* *Hemithylaca viridiventris*, Cab. et Hein. Mus. Hein. Theil iii. p. 38.

Habitat. New Granada. Common in the neighbourhood of Bogota.

350. PYRRHOPHÆNA IODURA.

Trochilus iodurus, Sauc. in Mus. Hein.
* *Saucerottia iodura*, Reichenb. Aufz. der Col. p. 8.
* *Chlorestes iodurus*, Reichenb. Troch. Enum. p. 4, pl. 698. figs. 4560-61.
* *Hemithylaca iodura*, Cab. et Hein. Mus. Hein. Theil iii. p. 39.
* *Trochilus Aglaiæ*, Bourc. Ann. Soc. Sci. Phys. et Nat. Lyon, 1846, p. 329? Id. Rev. Zool. 1846, p. 318?

* *Polytmus Aglaiæ*, Gray & Mitch. Gen. of Birds, vol. i. p. 109, *Polytmus*, sp. 73?
* *Amazilius aglaiæ*, Bonap. Consp. Gen. Av. tom. i. p. 71, *Amazilius*, sp. 11?
* *Saucerottia Aglaiæ*, Reichenb. Aufz. der Col. p. 8?
* *Chlorestes Aglaiae*, Reichenb. Troch. Enum. p. 4?
* *Hemithylaca Aglaiae*, Cab. et Hein. Mus. Hein. Theil iii. p. 38, note?

Habitat. New Granada?

The specimen named *iodura* in the Museum of M. Heine at Halberstadt is different from the bird so called in the Museum at Berlin. The former has a glittering crown and light-lilac shining tail-feathers; while the latter has a dull-coloured crown, and the tail so nearly resembling that of *P. viridigaster*, that I have no doubt of its being a young bird, or a female of that species. On the other hand, I think M. Heine's bird is a distinct species, and I have therefore retained it under the term *iodura*.

351. PYRRHOPHÆNA CYANURA.

Amazilia cyanura, *Gould* . . . . . . Vol. V. Pl. CCCXV.
* *Hemithylaca cyanura*, Cab. et Hein. Mus. Hein. p. 38, note.

Habitat. Pacific side of Nicaragua; Realjo?

The *Erythronotæ* are nearly allied to the *Pyrrhophænæ*. As in that genus, the sexes are alike in their colouring. The species are much more circumscribed in their habitat, being almost confined to Venezuela and the countries immediately adjoining.

The oldest-known species, which I have called *antiqua*, appears to be subject to much variation in its colour and markings; but, as I stated in my account of that species, I have questioned the propriety of their separation until we are better acquainted with them.

Genus ERYTHRONOTA, *Gould*.

('Ερυθρὸς, ruber, et νῶτος, dorsum.)

Generic characters.
Male.—*Bill* longer than the head, nearly straight or very slightly curved; *wings* moderately long; *tail* rather short and slightly forked; *tarsi* clothed; *hind-toe* rather shorter than the middle one; *claws* short; *throat* and *chest* lively green.

Female. Similar in colour.

352. ERYTHRONOTA ANTIQUA . . . . Vol. V. Pl. CCCXVI.

Ornismya erythronotos, Less.
——— *erythronotus*, Less.
Polytmus erythronotus, Gray & Mitch.
Saucerottia erythronota, Bonap., Reich.
Trochilus erythronotus, Jard.

* *Cynanthus erythronotus*, Jard. Nat. Lib. Humming Birds, vol. II. p. 148.
* *Chlorestes erythronotus*, Reichenb. Troch. Enum. p. 4, pl. 799. figs. 4562-63.
* *Hemithylaca erythronota*, Cab. et Hein. Mus. Hein. Theil iiL p. 37.

Habitat. Trinidad, Tobago, and Venezuela.

The birds from Tobago are very much larger than those from Trinidad; and some of the specimens from Venezuela have the under tail-coverts wholly chestnut; I should have considered this indicative of another species, had I not found a similar variation in specimens from Trinidad.

353. ERYTHRONOTA FELICIÆ . . . . Vol. V. Pl. CCCXVII.

Ornismya Feliciæ, Less.
Saucerottia Feliciae, Reichenb.
———— *felicia*, Bonap.
Chlorestes Feliciae, Reichenb.
Hemithylaca Feliciae, Cab. et Hein.
Trochilus Emile.

Habitat. Venezuela.

The *O. Feliciæ* is admitted by the French Trochilidists to be distinct from *O. antiqua*, and as such I have kept it; at the same time it is extremely difficult to distinguish one from the other. In size they are as near alike as possible; but the former has a bluer tail, and the back and upper surface destitute of the fiery red colouring observable in many specimens, but not in all, of the *O. antiqua*; the under tail-coverts, too, are frequently stained with violet.

I have a specimen of this bird, presented to me by M. Emile Parzudaki, of Paris, with the name of *Emile* attached. The bird was killed by him during his visit to South America.

354. ERYTHRONOTA EDWARDI . . . Vol. V. Pl. CCCXVIII.

Trochilus Edward, Delatt. et Bourc.
Polytmus Edwardsii, Gray & Mitch.
Amazilius edward, Bonap.
Thaumantias edward, Bonap.
Saucerottia Edwardsii, Reich.
* *Chlorestes Edwardsii*, Reichenb. Troch. Enum. p. 4, pl. 698. figs. 4558-60.
* *Hemithylaca Edwardi*, Cab. et Hein. Mus. Hein. Theil iii. p. 37, note.

Habitat. Panama, Costa Rica, and Veragua.

I have specimens of this bird killed by Mr. Bridges near David, at an elevation of from 8000 to 10,000 feet, according to the label attached.

355. ERYTHRONOTA NIVEIVENTRIS, *Gould* Vol. V. Pl. CCCXIX.

Trochilus (———?) *niveoventer*, Gould.

Thaumantias niveiventer, Bonap.
Saucerottia niveiventer, Reich.
* *Chlorestes niveiventris*, Reichenb. Troch. Enum. p. 4, pl. 700.
figs. 4566–67.
* *Hemithylaca niveiventris*, Cab. et Hein. Mus. Hein. Theil iii.
p. 37.
Habitat. Panama and Veragua.

856. ERYTHRONOTA ELEGANS, *Gould* . . Vol. V. Pl. CCCXX.
Erythronota elegans, Gould, Proc. of Zool. Soc. pt. xxviii. p. 307.
Habitat. Unknown.

This is a very elegant species, and quite distinct from every other known Humming-Bird. In its glittering light-green crown, throat, and chest it looks like a *Chlorostilbon*; but the form of its tail and some other characters ally it to the *Erythronotæ*, with which I have provisionally placed it.

The next natural section is that of *Saucerottia*, of which I am acquainted with three species distinguished from the *Erythronotæ* by their larger size, stouter bills, and by their more uniform dark-green colouring. All are confined to a comparatively limited area—namely, Costa Rica, Veragua, Panama, and the northern parts of New Granada.

Genus SAUCEROTTIA, *Bonap.*

857. SAUCEROTTIA TYPICA, *Bonap.*
Erythronota Saucerottei . . . . . . Vol. V. Pl. CCCXXI.
Trochilus Saucerottii, Bourc. et Delatt.
Polytmus Saucerottii, Gray & Mitch.
Saucerottia typica, Bonap.
Chlorestes typicus, Reichenb.
Hemithylaca Saucerottei, Cab. et Hein.
Habitat. New Granada.

858. SAUCEROTTIA SOPHIÆ.
Erythronota Sophiæ . . . . Vol. V. Pl. CCCXXII.
Trochilus Sophiæ, Bourc.
Polytmus Sophiæ, Gray & Mitch.
Amazilius sophiæ, Bonap.
Saucerottia Sophiae, Reich.
——— *sophiæ*, Sclat.
——— *sophia*, Bonap.
Chlorestes Sophiae, Reich.
Trochilus (——?) *caligatus*, Gould.
Hemithylaca Sophiae, Cab. et Hein.
——— *Hoffmanni*, Cab. et Hein.
Habitat. Costa Rica, Veragua, and New Granada.

359. SAUCEROTTIA WARSZEWICZI.

Hemithylaca Warszewiczi, Cab. et Hein. Mus. Hein. Theil iii. p. 38.

Habitat. Banks of the River Magdalena.

As the *S. Sophiæ* differs from the *S. typica* in the richer blue colouring of its upper and under tail-coverts and tail, so does this species differ from the *S. Sophiæ* in having the tail and its coverts both above and beneath of a still richer and more violet blue. It is also of smaller size; and the green of its under surface is different from that of both, being purer and deeper. The examples in my collection were obtained by M. Warszewicz on the banks of the Magdalena.

The bird to which M. Bourcier has given the specific name of *cyanifrons* requires separation from the last three species, as much or more than they do from their predecessors the *Erythronotæ*. M. Cabanis's generic name of *Hemithylaca* having been applied to this group as a whole, subsequently to those of *Saucerottia* and *Erythronota*, I must either place his name in the rank of a synonym or adopt it for the present species, the only one of this particular form.

Genus HEMITHYLACA, *Cab.*

360. HEMITHYLACA CYANIFRONS.

Saucerottia cyanifrons . . . . . . Vol. V. Pl. CCCXXIII.
Trochilus cyanifrons, Bourc.
Polytmus cyanifrons, Gray & Mitch.
Thalurania cyanifrons, Bonap.
Saucerottia cyanifrons, Bonap., Reichenb.
Chlorestes cyanifrons, Reichenb.
Hemithylaca cyanifrons, Cab. et Hein. Mus. Hein. Theil iii. p. 39.

Habitat. New Granada.

Somewhat allied to the genera *Hemithylaca* and *Erythronota* is the isolated form constituting my genus *Eupherusa*. The single species known is a native of Central America. Contrary to what occurs among the *Erythronotæ*, the sexes differ very considerably in their plumage; a fact of which I was not aware when my plate of the species was executed.

Genus EUPHERUSA, *Gould.*

(Εὖ, benè, feliciter; et φέρουσα, gestans.)

Male.—Bill nearly straight and longer than the head; wing rather long; tail rounded; tarsi clothed; feet small; hind toe rather shorter than the middle toe.
Female.—Unadorned.

361. EUPHERUSA EXIMIA . . . . . . Vol. V. Pl. CCCXXIV
Trochilus eximius, Delatt.

Saucerottia eximia, Reich., Bonap.
Amazilia eximia, Reichenb. Troch. Enum. p. 8, pl. 776. fig. 4802.

Habitat. Guatemala and Honduras?

Mr. Salvin states that "This is one of the commonest Humming-Birds of Coban, being found everywhere near the city. The ratio of the males to the females is as ten to four."—*Ibis*, vol. ii. p. 271.

The following is a correct description of the female of this species:—

Throat and all the under surface grey; sides of the neck and upper surface green; primaries purplish brown; secondaries deep buff, forming epaulets as in the male, but of lesser size.

The

Genus CHRYSURONIA, *Bonap.*,

is composed of six species, with pretty, golden tails. The females of most of them are strikingly different; for, although they all have the tail similarly coloured, they are destitute of brilliancy on any part of the body. All are inhabitants of the Andes in New Granada, Ecuador, and Peru, with the exception of the *C. Eliciæ*, which inhabits countries to the northward of Panama.

362. CHRYSURONIA ŒNONE . . . . . Vol. V. Pl. CCCXXV.

Ornismya Oenone, Less., Delatt.
Polytmus Œnone, Gray & Mitch.
Chrysuronia œnone, Bonap., Sclat.
——— *Oenone*, Reichenb.
Cynanthus Oenone, Jard. Nat. Lib. HummingBirds, vol. ii. p 149.
Chrysurisca Oenone, Cab. et Hein. Mus. Hein. Theil iii. p. 42.

Habitat. Venezuela and New Granada.

I find no difference between the birds from Venezuela and those from the neighbourhood of Bogota, except that the latter have rather longer bills, and the tail-feathers lighter and inclined to green.

363. CHRYSURONIA JOSEPHINÆ . . . Vol. V. Pl. CCCXXVI.

Ornismyia Josephinæ, Bourc. et Mulс.
Trochilus Josephinæ, Gray & Mitch.
Chrysuronia Josephinae, Reichenb., Bonap.
Chrysurisca Josephinae, Cab. et Hein. Mus. Hein. Theil iii. p. 42, note.

Habitat. The upper parts of the River Amazon, where specimens were procured by Mr. Bates.

I have two specimens of a bird of this form in my collection, one of which, procured in Paris, is labelled "*O. neera*, Less.;" they differ from every other species I possess. They are much larger than *C. Josephinæ*, and their tails are rich fiery bronze; their crowns greenish blue; all the under surface golden green; the under tail-coverts fiery bronze like the tail; and the blue of the crown extends further down the neck. I therefore retain the name of *nœra* for this bird. I have still another bird allied to *Josephinœ*, with a longer wing, a shorter tail, and a somewhat shorter bill; the colour of the crown is violet or purplish-blue instead of greenish-blue, and the tail, instead of being rich fiery bronze, is light greenish bronze. I cannot do otherwise than provisionally name this bird, and I therefore propose to call it *C. cœruleicapilla*.

364. CHRYSURONIA NÆRA.
Ornysmia nœra, Less., Delatt. et Less. Rev. Zool. 1839, p. 18.
Habitat. Unknown.

365. CHRYSURONIA CÆRULEICAPILLA, *Gould.*
Habitat. Unknown.

366. CHRYSURONIA HUMBOLDTI . . Vol. V. Pl. CCCXXVII.
Trochilus Humboldti, Bourc. et Muls.
Chrysuronia Humboldti, Reichenb., Bonap.
* *Chrysurisca Humboldti*, Cab. et Hein. Mus. Hein. Theil iii. p. 42, note.
Habitat. The banks of the River Miva in the province of Esmeraldas in Ecuador.

367. CHRYSURONIA ELICIÆ . . . . Vol. V. Pl. CCCXXVIII.
Trochilus Eliciæ, Bourc. et Muls.
Polytmus Eliciæ, Gray & Mitch.
Chrysuronia elicia, Bonap.
———— *Eliciae*, Reichenb., Bonap.
* *Chrysurisca Eliciae*, Cab. et Hein. Mus. Hein. Theil iii. p. 42.
Habitat. Guatemala, Costa Rica, and Veragua.

368. CHRYSURONIA CHRYSURA . . . Vol. V. Pl. CCCXXIX.
Ornismya chrysura, Less.
Chrysuronia chrysura, Bonap., Reichenb.
Polytmus chrysura, Gray & Mitch.
* *Phaëthornis? chrysurus*, Jard. Nat. Lib. Humming Birds, vol. ii. p. 152.
* *Chrysurisca chrysura*, Cab. et Hein. Mus. Hein. Theil iii. p. 42, note.
Habitat. Peru.

Rich and conspicuous blue is the prevailing tint in the genera *Eucephala* and *Hylocharis*, which may be considered as truly Brazilian, since most of the species are natives of that country, almost the only exception being the *E. Grayi*, which is found in the Andes. There is scarcely any section of the Trochilidæ less understood or more difficult to discriminate than the next six or eight species.

Genus EUCEPHALA, *Reichenb.*

369. EUCEPHALA GRAYI . . . . . . Vol. V. Pl. CCCXXX.
Trochilus Grayi, Delatt. et Boure.
Hylocharis Grayii, Gray & Mitch., Bonap.
Eucephala Grayi, Reichenb.
Sapphironia grayi, Bonap.
*Eucephala Grayi, Cab. et Hein. Mus. Hein. Theil iii. p. 43.

Habitat. Said to be Popayan in New Granada.

This is the largest species of the genus, and a very rare bird.

370. EUCEPHALA SMARAGDO-CÆRULEA, *Gould.*
Vol. V. Pl. CCCXXXI.
Augasma smaragdineum, Gould.

Habitat. Brazil, from Rio de Janeiro towards the interior.

371. EUCEPHALA CHLOROCEPHALA . . Vol. V. Pl. CCCXXXII.
Hylocharis chlorocephala, Boure.
——— *chlorocephalus*, Bonap.
Agyrtria chlorocephala, Reichenb.
Lepidopyga chlorocephala, Cab. et Hein.

Habitat. The environs of Guaranda in Ecuador, according to M. Bourcier.

372. EUCEPHALA CÆRULEO-LAVATA, *Gould.*
Vol. V. Pl. CCCXXXIII.
Habitat. South-eastern Brazil.

373. EUCEPHALA SCAPULATA, *Gould.*
Habitat. Supposed to be Cayenne.

Crown of the head, back of the neck, and lower part of the back very deep dull green; throat and chest glittering greenish blue, imperceptibly passing into the dull brownish black of the abdomen; under tail-coverts brown, with a wash of dull blue in the centre of each feather; a mark of blue on each side at the insertion of the wing, forming an indistinct band across the back; upper tail-coverts bronzy green; tail steely black, rather short for the size of the bird, and slightly forked; wings deep purplish brown; tarsi clothed with intermingled greyish-white and brown feathers; upper mandible black; basal half of the under mandible fleshy, the apical half black.

Total length 3¾ inches, bill ⅝, wing 2 1/16, tail 1⅜.

In the size of its body, it nearly equals the *Eucephala cæruleo-lavata*, but it differs from that and every other known species of this family of birds.

I have only seen a single example of this species.

374. EUCEPHALA HYPOCYANEA, *Gould* Vol. V. Pl. CCCXXXIV.
Habitat. Unknown, probably Brazil.

375. EUCEPHALA CÆRULEA . . . . Vol. V. Pl. CCCXXXV.
Trochilus cæruleus, Vieill.
Ornismya Audeberti, Less., Bourc.
Hylocharis cærulea, Gray.
——— *cæruleus*, Bonap.
Thaumatias cæruleus, Bonap.
Chlorestes coerulea, Reichenb.
**Chlorestes cæruleus*, Cab. et Hein. Mus. Hein. Theil iii. p. 46, note.
**Trochilus Audeberti*, Wied, Beitr. iv. p. 67.
**Cynanthus ? cæruleus*, Jard. Nat. Lib. Humming Birds, vol. ii. p. 147.
**Hylocharis Audeberti*, Burm. Th. Bras. tom. ii. p. 349.

Habitat. Eastern and Northern Brazil (Chamicuros, *Hauxwell*), the Guianas, Venezuela, Trinidad, and Tobago.

Specimens from all these localities are so much alike that it is impossible to consider them otherwise than as one and the same species; but I may remark that those from Venezuela have the blue mark on the chin much less apparent than those from Cayenne, Trinidad, and Eastern Brazil. My Chamicuros specimen also has this colour but faintly indicated, and the tail somewhat larger.

376. EUCEPHALA CYANOGENYS.
**Trochilus cyanogenys*, Wied, Beitr. iv. p. 10; Jard. Nat. Lib. Humming Birds, vol. ii. p. 89; Burm. Their. Bras. tom. ii. p. 350.
*Ornismya Wiedii, Less. Supp. Hist. Nat. des Ois.-mou. p. 150, pl. 26.
**Cynanthus cyanogenys*, Jard. Nat. Lib. Humming Birds, vol. ii. p. 148.
**Hylocharis cyanogenys*, Gray, Gen. of Birds, vol. i. p. 115, *Hylocharis*, sp. 40.
**Sauceroittia cyanogenys*, Bonap. Gen. Av. tom. l. p. 77, *Saucerottia*, sp. 3.
*Hylocharis wiedi, Bonap. Rev. et Mag. de Zool. 1854, p. 255.
**Chlorestes cyanogenys*, Reichenb. Aufz. der Col. p. 7; Id. Troch. Enum. p. 4, pl. 692. figs. 4536-37; Cab. et Hein. Mus. Hein. Theil iii. p. 46.

Habitat. Brazil.

The single example of this bird procured by Prince Maximilian of Wied is the only one that has been seen. It is very closely allied to, but smaller than, *E. cærulea*.

Genus PANTERPE, *Cab.*

This generic name has been proposed by M. Cabanis for the beautiful new bird discovered by Dr. Hoffmann in Costa Rica, of which I believe only a single example was obtained. Nothing is known with regard to the colouring of the sexes.

377. PANTERPE INSIGNIS, *Cab.* . . . . Vol. V. Pl. CCCXXXVI.
Habitat. Costa Rica.

The member or members, as the case may be, of the

Genus JULIAMYIA, *Bonap.*,

stand quite alone and apart from all the other small Humming-Birds, and bear the same relationship to the *Eucephala* that the *Sphenoproctus Pampa* does to the *Campylopteri*. Some of the specimens of this form have brilliantly glittering crowns; in others this part of the head is dull-coloured; while the plumage of the body is alike in all.

These differences have sadly perplexed me for many years; but after a very careful and minute examination of a great number of examples from various localities, I believe I shall be right in regarding the brilliantly coronetted bird as distinct from its dull-crowned ally, and in adopting the name of *Feliciana* of Lesson, believing that his description has reference to it.

378. JULIAMYIA TYPICA . . . . . Vol. V. Pl. CCCXXXVII.
Ornismyia Julie, Buurc.
Ornismya Juliæ, Buurc.
Hylocharis Juliæ, Gray & Mitch.
Damophila Julia, Reichenb.
Juliamyia typica, Bonap.
* *Cœligena Juliae*, Reichenb. Troch. Enum. p. 3, pl. 681. figs. 4494–95, and pl. 763. fig. 4767.
* *Damophila Juliae*. Cab. et Hein. Mus. Hein. Theil iii. p. 40.
Habitat. New Granada.

379. JULIAMYIA FELICIANA.
* *Ornismya Feliciana*, Less. Rev. Zool. 1844, p. 493.
* *Hylocharis Feliciana*, Gray & Mitch. Gen. of Birds, vol. I. p. 114, *Hylocharis*, sp. 27.

Habitat. Ecuador.

Mr. Fraser states that at Bababoyo this species is "not very common, and only found in the deep bush, where it feeds on the tops of good-sized trees," and that in Esmeraldas it was "taken catching flies among the Cacao plantations. In October common everywhere; in December rare." "Irides hazel; upper mandible black; lower red, with black tip."—*Proc. of Zool. Soc.* 1860, pp. 283, 296.

The Mexican genus *Circe* comprehends, as far as our present knowledge extends, only two species. Although their tails are somewhat short, they are composed of broad and ample feathers, all of which are tipped with grey; in this respect they remind us of the *Chlorolampis Caniveti* and its allies.

Genus CIRCE, *Gould.*
(Κίρκη, Circe.)

Generic characters.
Male.—Bill rather longer than the head, slightly curved and

tapering towards the tip; *wings* rather long; *primaries* rigid; *tail*
rather short, and slightly forked, the feathers broad; *tarsi* clothed;
feet small; *hind toe* short; *throat* slightly luminous.
Female.—Very dull in colour.

The six middle tail-feathers of both species are edged with brown,
as in *Cœliveti.*

380. CIRCE LATIROSTRIS . . . . Vol. V. Pl. CCCXXXVIII.

Trochilus latirostris, Swains.
——— *Circe,* Bourc.
Sapphironia circe, Bonap.
Trochilus Schimperi, Saucer. MS.
* *Ornismya Lessoni,* Delatt. Rev. Zool. 1839, p. 15 (female).
* *Trochilus lazulus,* Licht. in Mus. of Berlin.
* *Cyanophaia lazula,* Reichenb. Aufz. der Col. p. 10.
* *Hylocharis lazula,* Reichenb. Troch. Enum. p. 8, pl. 770. figs.
4783–84.
* ——— *Doubledayi,* Cab. et Hein. Mus. Hein. Theil iii. p. 44.

Habitat. The table-lands of Mexico.

My late friend Dr. Saucerotte gave me the type-specimen of his
Trochilus Schimperi with the name attached, by which means I am
able to state that it is identical with the present bird, and not with
the *Trochilus lazulus* of Vieillot as supposed by Dr. Reichenbach
(see his Aufz. der Col. p. 21).

381. CIRCE DOUBLEDAYI . . . . . Vol. V. Pl. CCCXXXIX.

Trochilus Doubledayi, Bourc.
Thaumatias doubledayi, Bonap.
Cyanophaia Doubledayi, Reichenb.
Hylocharis Doubledayi, Gray & Mitch.
Sapphironia doubledayi, Bonap.
Trochilus Lereboullei, Sauc. MS.

Habitat. Mexico; locality Chimantla, according to Dr. Saucerotte.

Genus PHÆOPTILA, *Gould.*

(Φαιός, obscurus, et πτίλον, pluma.)

Generic characters.

Male.—Bill longer than the head, fleshy at the base, and slightly
arched; *wings* of medium length; *tail* the same, and slightly forked;
feet rather stout; *hind toe* and *nail* shorter than the middle toe and
nail.

382. PHÆOPTILA SORDIDA, *Gould* . . . Vol. V. Pl. CCCXL.

Cyanomyia? sordida, Gould.
——— *sordida,* Sclat.
Urunomitra sordida, Cab. et Hein.

Habitat. Oaxaca in Mexico.

There is a specimen in the Loddigesian collection which appears to be distinct from this bird; without figuring I will give a description, and propose for it the name of *Phæoptila zonura*.

383. PHÆOPTILA ZONURA, *Gould.*

Habitat. Bolanos in Mexico.

This bird, which is rather smaller and more delicately formed than *P. obscura*, has all the upper surface dull bronzy green; a stripe of greyish white over each eye; ear-coverts dusky; wings purplish brown; two centre tail-feathers bronzy green; the remainder bronzy green, crossed near the extremity with a broad band of blackish brown, beyond which the tips are greyish brown; all the under surface grey.

The single species of the

Genus DAMOPHILA, *Reichenb.*,

stands quite alone, no second member of the form having yet been discovered. Its native country is the Andes of New Granada and Ecuador, from both of which localities specimens are now before me. Those procured by Mr. Fraser in Esmeraldas differ a little from the specimens commonly sent in collections from Bogota, in having the centre of the throat greyish brown instead of black, and the two centre tail-feathers reddish purple instead of bronzy green: they must not, however, I think, be regarded as other than local varieties. Mr. Fraser's specimens appear not to be fully adult.

384. DAMOPHILA AMABILIS, *Gould* . . . Vol. V. Pl. CCCXLI.

Trochilus (——— ?) *amabilis*, Gould.
Damophila amabilis, Reichenb.
Juliamyia amabilis, Bonap.
* *Coeligena amabilis*, Reichenb. Troch. Enum. p. 3, pl. 681. figs. 4496–97.
* *Damophila amabilis*, Cab. et Hein. Mus. Hein. Theil iii. p. 40.

Habitat. New Granada and Ecuador.

I am not quite certain that we are acquainted with the true female of this bird, but I believe my representation of that sex to be correct; if so, the sexes are very dissimilar in colouring, and in this respect are closely allied to the *Eucephalæ*.

Boié, one of the most philosophical of modern ornithologists, proposed the generic term of *Hylocharis* for the *Trochilus sapphirinus* of Gmelin; and I have much pleasure in adopting this name, as well as several others proposed by him. The

Genus HYLOCHARIS, *Boié*,

is composed of three species, all of which are natives of Brazil, to which country they are mainly confined. They are all very pretty, if not showy, species; and one of them is among the oldest-known members of the entire family, as will be seen on reference to the synonyms; brilliant blue is the prevailing colour of the males, while the opposite sex is but plainly attired.

385. HYLOCHARIS SAPPHIRINA . . . . Vol. V. Pl. CCCXLII.

Trochilus sapphirinus, Gmel., Lath., Shaw, Jard.
Ornismya sapphirina, Less.
Hylocharis sapphirina, Gray & Mitch.
——————— *sapphirinus*, Bonap.
Trochilus fulvifrons, Lath.
* *Trochilus sapphirinus*, Vieill. Ency. Méth. Orn. part ii. p. 570; Licht. Doubl. p. 14.
* ——————— *latirostris*, Wied, Beit. iv. p. 64.
* *Cynanthus sapphirinus*, Jard. Nat. Lib. Humming Birds, vol. ii. p. 147.
* *Hylocharis latirostris*, Reichenb. Aufz. der Col. p. 10.
* *Sapphironia sapphirina*, Bonap. Rev. et Mag. de Zool. 1854, p. 256.
* *Hylocharis sapphirina*, Reichenb. Troch. Enum. p. 7, pl. 769. figs. 4780–82; Burm. Th. Bras. tom. ii. p. 346; Cab. et Hein. Mus. Hein. Theil iii. p. 43.

Habitat. Brazil.

This species arrives in the neighbourhood of Rio de Janeiro in July, and departs again in November. I have also seen specimens from Bahia, Para, and other parts on the Lower Amazon.

386. HYLOCHARIS LACTEA . . . . . Vol. V. Pl. CCCXLIII.

Ornismya sapphirina, Less.
——————— *lactea*, Less.
Hylocharis lactea, Gray & Mitch., Bonap.
Cyanochloris lactea, Reich.
Sapphironia lactea, Bonap.
* *Trochilus sapphirinus*, Wied, Beit. iv. p. 61.
* ——————— *lazulinus*, Licht. in Mus. of Berlin.
* *Hylocharis lactea*, Cab. et Hein. Mus. Hein. Theil iii. p. 43, note; Reich. Troch. Euum. p. 6, pl. 773. figs. 4788–91.

Habitat. Eastern and Northern Brazil.

387. HYLOCHARIS CYANEA . . . . . Vol. V. Pl. CCCXLIV.

Trochilus cyaneus, Vieill., Jard.
Ornismya cyanea, Less.
Hylocharis cyanea, Gray & Mitch.
Thaumatias cyaneus, Bonap.
* *Trochilus azureus*, Licht. Doubl. p. 14.
* *Hylocharis cyanea*, Reichenb. Aufz. der Col. p. 10; Id. Troch. Enum. p. 7, pl. 768. figs. 4777–79; Cab. et Hein. Mus. Hein. Theil iii. p. 44.
* *Ornismya bicolor*, Less. Hist. Nat. des Ois.-mou. p. 161, pl. 49, 50?? Id. Traité d'Orn. p. 280?? Id. Les Troch. p. 58, pl. 16??

Habitat. Eastern Brazil from Rio Janeiro to Bahia, where it is stationary.

Next to *Hylocharis* I place the two species of the

Genus SAPPHIRONIA, *Bonap.*;

for although one of them has a glittering green breast, I consider it to be more nearly allied to that form than to the *Chlorostilbones* and their allies. They are very elegant in form, and I believe that the females of both species differ very considerably from the males. I consider this genus to be a very natural one; yet, strange to say, one of the species (*S. cæruleigularis*) has a blue breast, while the other is wholly green.

Both are natives of the Andes—one in Central America, the other in New Granada.

388. SAPPHIRONIA GOUDOTI . . . . Vol. V. Pl. CCCXLV.

Trochilus Goudoti, Bourc.
Saucerottia goudoti, Bonap.
Polytmus Goudoti, Gray & Mitch.
Chalybura Goudoti, Reichenb.
Hylocharis goudoti, Bonap., Sclat.
*Agyrtria Goudoti, Reichenb. Troch. Enum. p. 7, pl. 763. figs. 4765-66.
**Lepidopyga Goudoti*, Cab. et Hein. Mus. Hein. Theil iii. p. 40, note.

Habitat. New Granada.

389. SAPPHIRONIA CÆRULEIGULARIS, *Gould.*

Vol. V. Pl. CCCXLVI.

Trochilus (—?) *cæruleogularis*, Gould.
——— *Duchassaigni*, Bourc.
Thalurania Calina, Bourc.
Cyanochloris cæruleigularis, Reichenb.
Sapphironia cæruleigularis, Bonap.
Hylocharis (?) *cæruleigularis*, Sclat.
**Agyrtria cæruleigularis*, Reichenb. Troch. Enum. p. 7, pl. 764. figs. 4768-69.
* *Trochilus cyanomelas*, Licht. In Mus. of Berlin.
* *Cyanophaia cærulescens*, "Lodd.," Reichenb. in Mus. Heinean.
* *Hylocharis cærulescens*, Reichenb. Troch. Enum. p. 8, pl. 770. fig. 4785.
* *Lepidopyga cæruleigularis*, Cab. et Hein. Mus. Hein. Theil iii. p. 40.

Habitat. Costa Rica and Panama.

I now proceed to the genus

SPORADINUS, *Bonap.*,

the members of which are confined to the West India Islands. They are very elegant in form, and are of somewhat large size when compared with those which precede and follow them. They have deeply-forked tails, and the under surfaces of the males are brilliantly coloured; the females, on the other hand, are very sombrely attired.

390. SPORADINUS ELEGANS . . . . Vol. V. Pl. CCCXLVII.
Trochilus elegans, Aud. et Vieill.
Ornismya Swainsonii, Less.
Hylocharis elegans, Gray & Mitch.
Lampornis elegans, Bonap.
Riccordia elegans, Reichenb.
Sporadinus elegans, Bonap.
* *Trochilus Swainsoni*, Jard. Nat. Lib. Humm. Birds, vol. ii. p. 88.
* *Chlorestes elegans*, Reichenb. Trnch. Enum. p. 4, pl. 704. f. 4587.
* *Sporadinus elegans*, Cab. et Hein. Mus. Hein. Theil iii. p. 25.
Habitat. The Island of St. Domingo.

391. SPORADINUS RICORDI . . . . Vol. V. Pl. CCCXLVIII.
Orthorhynchus Ricordi, De la Sag.
Ornismya Parzudhaki, Less.
Sporadinus ricordi, Bonap.
Hylocharis Ricordi, Gray & Mitch.
Trochilus ricordi, Gerv., Bonap.
Riccordia Raimondi, Reichenb.
Chlorestes riccordi, Gundl.
* *Chlorestes Raimondii*, Reichenb. Troch. Enum. p. 4, pl. 704. figs. 4584–86.
* *Sporadinus Ricordi*, Cab. et Hein. Mus. Hein. Theil iii. p. 25, note.
Habitat. The Island of Cuba.

392. SPORADINUS? MAUGÆI . . . . Vol. V. Pl. CCCXLIX.
Mellisuga Surinamensis pectore cæruleo, Briss.
Trochilus Maugæus, Vieill.
Ornismya Maugæus, Less.
Trochilus Ourissia, auct.?
Habitat. Porto Rico.

This bird differs somewhat in form from the two preceding species, and it may be found necessary to institute a separate genus for its reception. It is a very rare bird, and the two specimens, male and female, in the collection at the Jardin des Plantes at Paris are the only ones I have ever seen.

The little glittering-green Humming-Birds forming the genera *Chlorolampis*, *Chloristilbon*, and *Panychlora*, are very widely spread over the temperate and warmer parts of the South American continent, being found along the whole course of the great Andean range, from Mexico on the north to Bolivia on the south; they also inhabit Brazil, Cayenne, the Guianas, Trinidad, and Venezuela. The sexes differ very considerably in colour in nearly every species, the males being clothed in a metallic covering, while that of the females is soft in texture and sombre in hue.

The members of the

GENUS CHLOROLAMPIS, *Cab.*,

as restricted by me, are distinguished for their deeply forked tails, most of the feathers of which are singularly tipped with dull grey. They range over a great part of Mexico, Guatemala, Costa Rica, Veragua, and Panama, where they apparently terminate, as I have never received examples from any part southward of the Isthmus. The finest species of the genus is the *C. auriceps*, a bird of very elegant form, and having a deeply forked tail; most nearly allied to this is the *C. Caniveti*: both these birds are from Mexico. Following these is the little bird I have named *C. Osberti*, after Mr. Osbert Salvin; a fourth, from Costa Rica, has been described by Dr. Cabanis as *C. Salvini*, after the same gentleman. Very considerable difference occurs in the sexes, but more in colour than in form; for the females have the tail forked like that of the males, but to a much less extent.

393. CHLOROLAMPIS AURICEPS.

Chlorostilbon auriceps, *Gould* . . . . Vol. V. Pl. CCCL.
Trochilus (— ?) *auriceps*, Gould.
* *Sporadinus auriceps*, Bonap. Rev. et Mag. de Zool. 1854, p. 255.
* *Trochilus modestus*, Licht. in Mus. of Berlin?
* *Chlorolampis auriceps*, Cab. et Hein. Mus. Hein. p. 48, note.
Habitat. Mexico: locality unknown.

394. CHLOROLAMPIS CANIVETI.

Chlorostilbon Caniveti . . . . Vol. V. Pl. CCCLI.
Ormismya Canivetii, Less.
Thaumatias caniveti, Bonap.
Riccordia Caniveti, Reichenb.
Sporadinus caniveti, Bonap.
Hyluchanis Caniveti, Gray & Mitch.
Chlorostilbon Caniveti, Sclat. & Salv.
* *Chlorestis Caniveti*, Troch. Enum. p. 4, pl. 703. figs. 4581–83.
* *Chlorolampis Caniveti*, Cab. et Hein. Mus. Hein. Theil iii. p. 47, note.
Habitat. Southern Mexico and Guatemala?

395. CHLOROLAMPIS OSBERTI, *Gould.*

Chlorostilbon Osberti, *Gould* . . . . Vol. V. Pl. CCCLII.
Habitat. The neighbourhood of Dueñas in Guatemala.

396. CHLORLAMPIS SALVINI, *Cab.*

Chlorolampis Salvini, Cab. et Hein. Mus. Hein. Theil iii. p. 48.
Habitat. Costa Rica according to Dr. Cabanis.

The *C. Salvini* is nearly allied to *C. Osberti* and to *C. Caniveti*, but I believe it to be distinct from both. The freshly moulted adult males have their four central tail-feathers tipped with bronzy-green; but this colour appears to fade upon exposure to light, leaving the tail nearly black. I believe this bird is also found at Panama.

Genus CHLOROSTILBON, *Gould.*

(Χλωρός, viridis, et στίλβω, corusco.)

Under this generic appellation, for a form of which I always intended the *C. prasinus* to be the type, I have figured the whole of the little green Humming-Birds; but I now see the necessity of subdividing them; I shall therefore restrict the term to the following species—*angustipennis, Haeberlini, Phaëthon, auriventris, prasinus, Atala, brevicaudatus, Napensis, Peruanus, Daphne,* and *chrysogaster,* and adopt Dr. Cabanis's genus *Panychlora* for *Aliciæ, euchloris, Poortmanni,* and *stenura.*

397. CHLOROSTILBON ANGUSTIPENNIS . Vol. V. Pl. CCCLIII.

Trochilus angustipennis, Fras.
Hylocharis angustipennis, Gray & Mitch.

Habitat. Panama and New Granada.

398. CHLOROSTILBON HAEBERLINI.

Chlorolampis chrysogaster, Cab. et Hein. Mus. Hein. Theil iii.
p. 47.
Trochilus Haeberlinii, Licht. in Mus. of Berlin.
Chlorestes Haeberlinii, Reichenb. Aufz. der Col. p. 7; Id. Troch.
Enum. p. 4, pl. 703. fig. 4578-80.
Chlorolampis Haeberlini, Cab. et Hein. Mus. Hein. Theil iii.
p. 48, note.

Habitat. Carthagena.

I have had the original of *C. Haeberlini* sent to me from Berlin, and I find it to be a very elegantly formed bird, nearly allied to, but quite distinct from, *C. angustipennis.* It differs in having the glittering green of the under surface washed with blue, a shorter wing, and a still more deeply forked tail, the feathers of which are steely-green, and not so dark as in that species. It is said to be from Carthagena.

399. CHLOROSTILBON PHAËTHON . . . Vol. V. Pl. CCCLIV.

Trochilus Phaëthon, Bourc.
——— *Phæton,* Gray & Mitch.
Chlorestes Phaëthon, Reich.
Hylocharis phaëton, Bonap.
——— *similis,* Bonap.
Chlorolampis Phaëthon, Cab. et Hein.
Trochilus flavifrons, Gould.
——— *metallicus,* Gould.
Trochilus similis, Bourc.?

Habitat. Bolivia, Southern Brazil, and La Plata.

Since writing my account of this species, in which I expressed my belief that the *Ornismyia auriventris* of D'Orbigny and Lafresnaye was identical with it, I have carefully re-examined my specimens from the above-named countries, together with an example collected by

Mr. Bridges, and I am now inclined to believe the *O. aureiventris* to
be distinct; but as it merely differs in being of smaller size in all its
admeasurements, it will not be necessary for me to figure it.

400. CHLOROSTILBON AUREIVENTRIS.

Ornismyia aureiventris, D'Orb. et Lafresn.
Hylocharis aureiventris, Bonap., Rev. et Mag. de Zool. 1854,
  p. 255.

Habitat. Bolivia and Peru.

401. CHLOROSTILBON PRASINUS . . . . Vol. V. Pl. CCCLV.

* *Trochilus Pucherani*, Bourc. et Muls. Rev. Zool. 1846, tom li.
  p. 271.
* *Hylocharis pucherani*, Bonap. Rev. et Mag. de Zool. 1854,
  p. 255.
* *Chlorestes Pucherani*, Reichenb. Aufz. der Col. p. 7 ; Id. Troch.
  Enum. p. 4, pl. 755. fig. 4796.
* *Trochilus nitidissimus*, Licht. in Mus. of Berlin.
* *Hylocharis prasina*, Burm. Th. Bras. tom. ii. p. 350.
* *Chlorestes nitidissimus*, Reichenb. Aufz. der Col. p. 7 ; Id. Troch.
  Enum. p. 4, pl. 693. figs. 4538–39.
* *Trochilus lamprus*, "Natt." in Mus. of Munich.
* *Chlorostilbon nitidissimus*, Cab. et Hein. Mus. Hein. Theil iii.
  p. 47.
* *Ornismya Galathea*, Bourc. et Muls. in Mus. of Paris.
* *Trochilus viridissimus*, Linn. in Mus. of Berlin (young).

In my account of this species I stated that, owing to its being
impossible to determine to what bird Lesson had given the name of
prasinus, I should apply it to the one generally known by that
term among collectors—the bird so common in the neighbourhood of
Rio de Janeiro, Minas Geraes, &c. From Dr. Cabanis we learn
that it has been named *Trochilus nitidissimus* by Lichtenstein in
the Museum of Berlin, and *Trochilus lamprus*, Natt. in the Museum
of Munich; but had either of these names been published to the
world before Dr. Cabanis included it in his 'Museum Heineanum'
under the name of *Chlorostilbon nitidissimus*? If not, and *prasinus* be rejected, that term must certainly give place to M. Bourcier's previously published one of *Pucherani*, which I find, from the
type specimen now before me, was given to a young male of this
species. Refer to my account of this species, and of *C. Atala*.

402. CHLOROSTILBON IGNEUS, *Gould.*

Habitat. Supposed to be the neighbourhood of Para.

Crown of the head glittering orange ; back of the neck and upper
surface fiery orange, becoming more intense on the wing-coverts ;
throat and chest glittering bluish green, gradually passing into the
fiery orange of the flanks and abdomen ; under tail-coverts green,
tinged with orange ; wings purplish brown ; tail purplish black ; bill
fleshy red at the base, gradually passing into the black of the tip.

This bird is about the same size as *C. prasinus*, but differs from that species in the fiery colouring above described, and in the tail being purplish- instead of steel-black.

This is the bird mentioned in my account of *C. prasinus* as having been sent to me by Mr. Reeves, of Rio de Janeiro. It is one of the most beautiful species of the family.

403. CHLOROSTILBON ATALA . . . . Vol. V. Pl. CCCLVI.

Ornismya Atala, Less.?
——— *prasina*, Less.?
Hylocharis atala, Gray & Mitch.
Saucerottia Atala, Reichenb., Bomp.
Trochilus Mellisugus, Linn.
*Chlorestes Atala, Reichenbach, Trochil. Enum. p. 4, pl. 700. fig. 4568.

Habitat. The Island of Trinidad and Venezuela.

I find that Venezuelan specimens differ a little from those of Trinidad, the green of the upper and under surface being more golden; still I have no doubt of their being identical.

404. CHLOROSTILBON DAPHNE.

Trochilus Daphne, Bourc.

Habitat. Peru.

I consider this to be a distinct species: it is very nearly allied to the Cayenne bird *C. Atala* of this work; but it has a more square tail, with the green of the chest strongly tinged with blue. I have M. Bourcier's type, which is labelled 'Voyage de Castelnau, Pampas del Sacramento.'

405. CHLOROSTILBON PERUANUS, *Gould*.

Habitat. Peru.

Bill black; crown, throat, and all the under surface glittering orange-green; upper surface bronzy green; wings brown; tail purplish black.

Total length 3¼ inches, bill ⅔, wing 1¼, tail 1¼.

This, one of the black-billed species, has even a more rounded tail than *C. Daphne*, from which it differs in its larger size and in having a longer bill, and especially in the glittering orange-green colouring of its breast, which in *C. Daphne* is blue. The *C. chrysogaster* has a somewhat forked steely-black tail; in other respects the two birds are very similar.

406. CHLOROSTILBON NAPENSIS, *Gould*.

Habitat. The banks of the River Napo.

This species is very similar to, but smaller than *C. Daphne*, has a still shorter tail, and the blue of the breast not so extended, or confined to the throat.

407. CHLOROSTILBON BREVICAUDATUS, *Gould.*

Habitat. Cayenne.

This bird is very similar to the *C. Atala* of Trinidad, has the same glittering green-coloured breast, but has a short and more truncate-formed tail, more so than *C. Daphne* or *C. Napensis.*

408. CHLOROSTILBON CHRYSOGASTER.

* *Trochilus chrysogaster*, Bourc. Ann. Soc. Sci. Phys. et Nat. Lyon, 1843, p. 40; Id. Rev. Zool. 1843, p. 101.
* *Hylocharis chrysogaster*, Gray & Mitch. Gen. of Birds, vol. i. p. 115, *Hylocharis*, sp. 43; Bonap. Consp. Gen. Av. tom. i. p. 74, *Hylocharis*, sp. 2.
* *Chlorestes chrysogaster*, Reichenb. Aufz. der Col. p. 7; Id. Troch. Enum. p. 4, pl. 693. figs. 4540–41.
* ——— *prasinus, fœm.* Id. ibid. pl. 755. fig. 4737 ?
* *Trochilus puber*, Siebold in Mus. Monac. (Cabanis).
* *Chlorolampis chrysogastra*, Cab. et Hein. Mus. Hein. Theil iii. p. 47.
* *Chlorostilbon melanorhynchus*, Gould in Proc. of Zool. Soc. part xxviii. p. 308 ?
* *Chlorolampis maragdina*, Cab. et Hein. Mus. Hein. Theil iii. p. 48 ?
* *Chlorostilbon atala*, Sclat. " List of Birds collected by Mr. Fraser at Pallatanga," in Proc. Zool. Soc. part xxvii. p. 145; Id. " List of Birds coll. by Mr. Fraser at Puellaro," in Proc. Zool. Soc. part xxviii. p. 94.

Habitat. New Granada and Ecuador.

In my description of *C. angustipennis* I stated that I considered the *Trochilus chrysogaster* of M. Bourcier to be identical with that species; but I have since more closely investigated the matter, and I now believe that this opinion was an erroneous one. I also believe that the *C. chrysogaster* and my *C. melanorhynchus* are one and the same bird; for I find little or no difference in the specimens from Panama, New Granada, and Ecuador. I further think it likely that the *C. smaragdina* of MM. Cabanis and Heine's 'Museum Heineanum' is also referable to it.

409. CHLOROSTILBON ASSIMILIS, *Lawr.*

Chlorostilbon assimilis, Lawr. Ann. of Lyc. of Nat. Hist. in New York, 1860, p. 292.

Habitat. Panama.

The following is Mr. Lawrence's description of his *C. assimilis* and his remarks on the species:—" The entire upper plumage is of bronze or dull golden-green; tail dark steel-blue; wings brownish purple; under plumage brilliant green, golden on the abdomen, and on the throat of a bluish green; under tail-coverts grass-green; a small white spot on the pleura; tibial feathers brown; bill and feet black.

" Length 3 inches, wing $1\frac{1}{8}$, tail $1\frac{1}{16}$, bill $\frac{8}{16}$."

"This species is somewhat like *C. melanorhynchus*, Gould (*chrysogaster*), but is smaller, and the crown is uniform in colour with the back, not brilliant. The latter species is also more golden on the abdomen, and has the tail less forked, with the feathers narrower."

410. CHLOROSTILBON NITENS, *Lawr.*

Chlorostilbon nitens, Lawr. Ann. Lyc. Nat. Hist. New York, April 22, 1861.

"*Habitat.* Venezuela.

"Front and crown golden yellowish-green, very brilliant; back and wing-coverts shining bronzed green, lower part of back and upper tail-coverts shining grass-green; under plumage brilliant green, of a bluish shade on the throat, and golden on the abdomen; tail steel-blue and forked; wings brownish-purple; turni clothed with blackish feathers; under tail-coverts bright grass-green; upper mandible black, the under yellowish for two-thirds its length, with the end black; feet black.

"Length 3 inches, wing $1\frac{2}{4}$, tail $1\frac{1}{4}$, bill $\frac{9}{8}$.

"Allied to *C. chrysogaster*, but is smaller, and has a very brilliant crown."

Among the smallest of the Trochilidæ are the members of the form to which Dr. Cabanis has given the name of *Panychlora*. They are all inhabitants of New Granada and Venezuela, and are known by the specific names of *Aliciæ*, *euchloris*, *Poortmanni*, and *stenura*. They are distinguished by their dull-green colouring, the extreme shortness of their tails, and by the great difference in the colouring of the sexes.

Genus PANYCHLORA, *Cab.*

The members of this genus form a very natural section among the little green Humming-Birds, very perceptible to those who have paid attention to this group of birds.

411. PANYCHLORA ALICIÆ.

Chlorostilbon Aliciæ . . . . . . Vol. V. Pl. CCCLVII.

Trochilus Alice, Boure.
Smaragdites Alice, Reichenb.
Chlorostilbon aliciæ, Bonap.
Chlorestes Aliciae, Reichenb. Troch. Enum. p. 4, pl. 754. figs. 4732-33.
Trochilus crypturus, Licht. in Mus. of Berlin.
Panychlora Aliciae, Cab. et Hein. Mus. Hein. p. 50, note.
——— aurata, Cab. et Hein. Mus. Hein. Theil iii. p. 50.
Smaragdites maculicollis, Reichenb. Aufz. der Col. p. 7.
Chlorestes maculicollis, Id. ibid. p. 24; Id. Troch. Enum. p. 4, pl. 694. figs. 4545-46; Cab. et Hein. Mus. Hein. Theil iii. p. 49, note.

Habitat. Venezuela and New Granada.

412. PANYCHLORA EUCHLORIS.

Smaragditis euchloris, Reichenb. Aufz. der Col. p. 7.
Chlorestes euchloris, Reichenb. ibid. p. 23; Id. Troch. Enum. p. 4, pl. 694. fig. 4544.

Habitat. New Granada?

There is a specimen in the Berlin Museum with a broken bill. In size it is rather larger than *Aliciæ*, the tail is more forked, and the two outer feathers more pointed; all the feathers have a purplish hue, as seen in *Poortmanni*, and the glittering feathers of the body are of a dull golden purplish green, as in that species.

413. PANYCHLORA STENURA, *Cab.*

* *Panychlora stenura*, Cab. et Hein. Mus. Hein. Theil iii. p. 56, note.
* *Chlorostilbon acuticaudus*, Gould in Proc. Zool. Soc. part xxviii. p. 308.

Habitat. Merida in New Granada.

This species is fully equal in size to the last, has a more lengthened bill, and the outer tail-feathers are much more pointed.

414. PANYCHLORA POORTMANNI.

Chlorostilbon Poortmanni . . . . . Vol. V. Pl. CCCLVIII.
Ornismya Poortmani, Bourc.
Hylocharis Poortmanni, Gray & Mitch., Bonap.
Chlorestes Poortmanni, Reichenb.
Chlorostilbon poortmani, Bonap.
* *Ornismya Esmeralda*, Less. in Mus. Heinean.
* *Smaragditis Esmeralda*, Reich. Auz. der Col. p. 7.
* *Chlorostilbon Esmeralda*, Reichenb. Troch. Enum. p. 4, pl. 694. figs. 4542–43.
* *Panychlora Poortmanni*, Cab. et Hein. Mus. Hein. Theil iii. p. 50.

Habitat. New Granada.

I shall close this account of the little green Humming-Birds with a description of the extraordinary species sent to me by Mr. Reeves of Rio de Janeiro, and which I have described in the 'Proceedings of the Zoological Society' as *Calliphlox? iridescens*. Its iridescent green colouring would indicate that it belongs to this section; while its comparatively small wings and short tail ally it to *Calliphlox*; but as it is not strictly referable to either genus, I propose for it a separate distinctive appellation, and provisionally placed here the

Genus SMARAGDOCHRYSIS, *Gould.*

(Σμάραγδοι, smaragdus, et χρῦσος, aurum.)

Generic characters.

Male.—*Bill* longer than the head, straight and slender; *wings* small, primaries narrow and rigid; *tail* of moderate size and deeply forked; *tarsi* clothed; *feet* small; *hind-toe* and *nail* nearly as long as the middle-toe and nail.

415. SMARAGDOCHRYSIS IRIDESCENS, *Gould* Vol. V. Pl. CCCLIX.
Calliphlox? iridescens, Gould in Proc. of Zool. Soc. part xxviii.
p. 310.
Habitat. The virgin forests of the interior of Brazil.

Genus PHLOGOPHILUS, *Gould.*
(Φλὸξ [φλογὸς], *nomen floræ*, et φίλος, amicus.)

Generic characters.

Male.—Bill straight; *wings* ample and rather rounded; *tarsi* long for a Humming-Bird, and bare; *tail* rather large and rounded; *hind toe* and *nail* shorter than the middle toe and nail.

The specimen from which the above characters were taken differs from every other known Humming-Bird in its more lengthened tarsi, and in the colouring of its rounded tail. The bird, which is immature, was received from the borders of the River Napo.

416. PHLOGOPHILUS HEMILEUCURUS, *Gould* Vol. V. Pl. CCCLX.
Phlogophilus hemileucurus, Gould in Proc. of Zool. Soc. part xxviii. p. 310.
Habitat. The banks of the River Napo?

In placing this bird at the end of my Monograph of the Trochilidæ, I do not wish it to be understood that this is its proper situation. I cannot imagine what the adult will be like, and consequently cannot tell to which genus of the family it is allied; but I believe, to *Aitelomyia.*

---

*Note.—*In the body of the work, Columbia has been given as the habitat of many of the species; but in this Introduction, Venezuela, New Granada, and Ecuador have been substituted, as the case required, for that more general term. A difference of opinion exists as to the correct spelling of New Granada,—some considering that it should be Grenada, and others Gronada; the latter has been adopted in this Introduction, while in the body of the folio work it is usually, if not always, spelt Grenada.

The Index to the specific names of Humming-Birds comprises every term of this kind with which I am acquainted. Among them are some which are not elsewhere mentioned in this Introduction; these are the specific appellations occurring in the works of the older and a few of the more modern authors, which I have found it quite impossible to ascertain to what birds they have been applied. It is but fair to state that the *Urolampra chloropogon* of Cabanis and Heine, and the *Chlorestes iolaimus* of Reichenbach, appear from the descriptions and figures to be good species; but, as I have not seen the typical examples, I am unable to speak positively respecting them; I shall, however, keep the subject of the Humming-Birds constantly before me, and, when desirable, place my remarks upon these, and any novelties that may occur, before the scientific world.

At page 11 I have stated that the Humming-Birds, like the Swifts, have ample wings and vast powers of flight. As this may appear contradictory to the remarks made on the wing-powers of *Selasphorus ruber* and *Trochilus Colubris* at p. 6, it will be as well to state, what I meant to convey is that their ample wings and bony structure is admirably adapted for sustaining them in the air for a considerable time, rather than for enabling them to take long flights from one country to another.

EXPLANATION OF THE ABBREVIATIONS,

AND

LIST OF THE AUTHORS AND WORKS REFERRED TO.

Albin.—*Albin, Natural History of Birds.*
Aud. Birds of Am.—*Audubon, Birds of America.*
Aud. Orn. Bio.—*Audubon, Ornithological Biography.*
Aud. Syn. Birds of Am.—*Audubon, Synopsis of the Birds of America.*
Aud. et Vieill.—*Audebert et Vieillot, Oiseaux dorés, ou à reflets métalliques.*
Azara Voy. dans l'Amér. Mér. Soc. edit.—*Azara, Voyage dans l'Amérique méridionale, Sonnini's edition.*
Banc. Hist. of Guiana.—*Bancroft, Natural History of Guiana.*
Bodd.—*Boddaert, Table des Planches enluminées d'Histoire Naturelle de M. D'Aubenton.*
Bois, in Oken's Isis.
Boiss. Mag. de Zool.—*Boissoneau, in Magasin de Zoologie.*
Boiss. Rev. Zool.—*Boissoneau, in Revue Zoologique.*
Bonap. Consp. Gen. Av.—*Bonaparte, Conspectus Generum Avium.*
Bonap. Consp. Troch. in Rev. et Mag. de Zool.—*Bonaparte, Conspectus Trochilorum, in Revue et Magasin de Zoologie.*
Bonap. Rev. et Mag. de Zool.—*Bonaparte, in Revue et Magasin de Zoologie.*
Bonap. Syn. Birds of U. States.—*Bonaparte, Synopsis of the Birds of the United States.*
Bonn. et Vieill.—*Bonnaterre and Vieillot, in Tableau Encyclopédie Méthodique, Part III. Ornithologie.*
Borowsk.—*Borowski, Vogel.*
Bourc. Ann. de la Soc. d'Agr. Hist. Nat. etc. de Lyon.—*Bourcier, in Annales de la Société d'Agriculture, Histoire Naturelle, etc. de Lyon.*
Bourc. Ann. Sci. Phys. et Nat. de Lyon.—*Bourcier, in Annales des Sciences Physiques et Naturelles de Lyon.*
Bourc. Compt. Rend. de l'Acad. des Sci.—*Bourcier, in Comptes Rendus de l'Académie des Sciences.*
Bourc. in Proc. Zool. Soc.—*Bourcier, in the Proceedings of the Zoological Society of London.*
Bourc. Rev. Zool.—*Bourcier, in Revue Zoologique.*
Bourc. et Mul's.—*Bourcier et Mulsant, in Revue Zoologique.*
Bourc. et Muls. Ann. de la Soc. Sci. de Lyon.—*Bourcier and Mulsant, in Annales des Sciences Physiques et Naturelles de Lyon.*
Bourc. et Muls. Ann. de l'Acad. Sci. Bell. Lett. et Arts de Lyon.—*Bourcier, in Annales de l'Académie des Sciences, Belles-Lettres et Arts de Lyon.*
Bourc. et Muls. Ann. de la Soc. Linn. de Lyon.—*Bourcier and Mulsant, in Annales de la Société Linnéenne de Lyon.*
Brandt, Icon. Av. Ross.—*Brandt, Descriptiones et Icones Animalium Rossicorum,* etc.
Bridges, Proc. of Zool. Soc.—*Bridges, in the Proceedings of the Zoological Society of London.*

Briss. Orn.—*Brisson, Ornithologie.*
Brown. Nat. Hist. of Jam.—*Browne, The Civil and Natural History of Jamaica.*
Bryant, List of Birds seen at the Bahamas.
Buff. Hist. Nat. des Ois.—*Buffon, Histoire Naturelle des Oiseaux.*
Buff. Sonn. edit.—*Sonnini's edit. of Buffon's Histoire Naturelle des Oiseaux.*
Buff. Pl. Enl.—*Buffon's Planches Enluminées.*
Burm. Th. Bras.—*Burmeister, Systematische Uebersicht der Thiere Brasiliens.*
Cab. and Cab. et Hein.—*Dr. Cabanis and Ferdinand Heine, Museum Heineanum.*
Cab. or Cabanis in Rich. Schomb. Reisen in Brit. Guian.—*Cabanis, in Schomburgk's Reisen in Britisch Guiana.*
Cabot in Proc. of Boston Soc. of Nat. Hist.—*Cabot, in the Proceedings of the Boston Society of Natural History.*
Cassin, Ill. Birds of California.—*Cassin, Illustrations of the Birds of California.*
Cuv. Règn. Anim.—*Cuvier, Règne Animal.*
Darwin, Zool. of Beagle.—*Darwin, The Zoology of the Voyage of H. M. S. Beagle, Part III. Birds, by John Gould.*
Da Silva, Maia Minerva Brasiliensis.
Delatt. Echo du Monde Savant.—*Delattre, in L'Echo du Monde Savant.*
Delatt. in Rev. Zool.—*Delattre, in Revue Zoologique.*
Delatt. et Bourc. Rev. Zool.—*Delattre and Bourcier, in Revue Zoologique.*
Delatt. et Less. Rev. Zool.—*Delattre and Lesson, in Revue Zoologique.*
De Longuem. Rev. Zool.—*De Longuemarre, in Revue Zoologique.*
Dev. Rev. et Mag. de Zool.—*Deville, in Revue et Magasin de Zoologie.*
D'Orb. Voy. dans l'Amér. Mérid. Ois.—*D'Orbigny, Voyage dans l'Amérique Méridionale: Oiseaux.*
D'Orb. et Lafres. Syn.—*D'Orbigny et Lafresnaye, Synopsis Avium.*
Drapiez, Dict. Class. d'Hist. Nat.—*Drapiez, in Le Dictionnaire Classique d'Histoire Naturelle.*
Dubus, Esquisses Orn.—*Dubus, Esquisses Ornithologiques.*
Dumont, Dict. Sci. Nat.—*Dumont de St. Croix, in Le Dictionnaire des Sciences Naturelles.*
Edwards, Birds, or Nat. Hist. of Birds.—*Edwards, Natural History of uncommon Birds.*
Edw. Glean. or Glean. of Nat. Hist.—*Edwards, Gleanings of Natural History.*
Ferm. Surinam.—*Fermin, Histoire Naturelle de Surinam.*
Fras. In Proc. Zool. Soc.—*Fraser, in the Proceedings of the Zoological Society of London.*
Gambel, Notes on Californian Birds.
Gambel, in Proc. of Acad. Sci. Philad.—*Gambel, in the Proceedings of the Academy of Sciences of Philadelphia.*
Gerv. Mag. de Zool.—*Gervais, in Le Magasin de Zoologie.*
Gmel. Linn. Syst. Nat.—*Gmelin's edition of Linnæus's Systema Naturæ.*
Gosse, Birds of Jamaica.
Gosse, Ill. Birds of Jam.—*Gosse, Illustrations of the Birds of Jamaica.*
Gould, in Ann. and Mag. Nat. Hist.—*Gould, in the Annals and Magazine of Natural History.*
Gould, in Jard. Contr. to Orn.—*Gould, in Jardine's Contributions to Ornithology.*
Gould, Proc. Zool. Soc.—*Gould, in the Proceedings of the Zoological Society of London.*
Gould, in Rep. Brit. Assoc.—*Gould, in the Report of the British Association.*
Gould, Zool. of Beagle.—*Gould, in Zoology of the Voyage of H. M. S. Beagle, Part III. Birds.*
Gray, Cat. of Gen. and Sub-gen. of Birds in Brit. Mus.—*G. R. Gray, Catalogue of the Genera and Sub-genera of Birds contained in the British Museum.*
Gray, List of Gen. of Birds.—*G. R. Gray, List of the Genera of Birds.*
Gray and Mitch.—*Gray and Mitchell, The Genera of Birds.*
Gundl. in Cab. Journ. fur Orn.—*Gundlach, in Cabanis's Journal für Ornithologie.*

Hill, Ann. and Mag. Nat. Hist.—*Hill, in the Annals and Magazine of Natural History.*
Jard. or Jardine in the Ann. and Mag. Nat. Hist.—*Jardine, in the Annals and Magazine of Natural History.*
Jard. Cont. to Orn.—*Jardine, Contributions to Ornithology.*
Jard. Nat. Lib. Humm. Birds.—*Jardine, Naturalist's Library, Humming-Birds.*
King in Proc. of Comm. of Sci. and Corr. of Zool. Soc.—*King, in the Proceedings of the Committee of Science and Correspondence of the Zoological Society of London.*
Klein Av. or Aves.—*Klein, Historia Avium Prodromus.*
Lath. Gen. Hist.—*Latham, General History of Birds.*
Lath. Gen. Syn.—*Latham, General Synopsis of Birds.*
Lath. Ind. Orn.—*Latham, Index Ornithologicus.*
Lawr. in Ann. Lyc. Nat. Hist. New York.—*Lawrence, in Annals of the Lyceum of Natural History in New York.*
Lembeye, Aves de l'Isle de Cuba.
Léry, Voyage au Brésil.
Less. Ann. Sci. Nat.—*Lesson, in Annales des Sciences Naturelles.*
Less. Col.—*Lesson, Histoire Naturelle des Colibris.*
Less. Echo du Monde Savant.—*Lesson, in l'Echo du Monde Savant.*
Less. Hist. Nat. des Col.—*Lesson, Histoire Naturelle des Colibris.*
Less. Hist. Nat. des Ois.-mou.—*Lesson, Histoire Naturelle des Oiseaux-mouches.*
Less. Ill. Zool.—*Lesson, Illustrations de Zoologie.*
Less. Ind. Gén. et Syn. des Ois. du Gen. Trochilus.—*Lesson, Index Général et Synoptique des Oiseaux du Genre Trochilus.*
Less. Les Troch.—*Lesson, Les Trochilidées.*
Less. Man. d'Orn.—*Lesson, Manuel d'Ornithologie.*
Less. Ois.-mou. Vélins.—*Lesson's unpublished additions to his Histoire Naturelle des Oiseaux-mouches.*
Less. Rev. Zool.—*Lesson, in Revue Zoologique.*
Less. Supp. Hist. Nat. des Ois.-mou.—*Lesson, Supplément à l'Histoire Naturelle des Oiseaux-mouches.*
Less. Tab. des Esp. des Ois.-mou.—*Lesson, Tableau des Espèces des Oiseaux-mouches.*
Less. Traité d'Orn.—*Lesson, Traité d'Ornithologie.*
Less. Voy. de la Coq.—*Lesson, in Le Voyage de la Coquille.*
Less. et Delatt. Rev. Zool.—*Lesson and De Lattre, in Revue Zoologique.*
Less. et Garn. Voy. de la Coq.—*Lesson and Garnot, in Le Voyage de la Coquille.*
Licht. Cat. of Birds in Mus. of Berlin.—*Lichtenstein, Catalogue of the Birds in the Museum of Berlin.*
Licht. Ermann. Verz. von Thier. und Pflanz.—*Atlas zu Ermann's Reise um die Welt.*
Licht. in Mus. Berlin.—*Lichtenstein, in the Berlin Museum.*
Licht. in Mus. Berol.—*Lichtenstein, in the Berlin Museum.*
Licht. Preis-Verz. Mexican. Thier. v. Deppe und Schiede.—*Lichtenstein, Preis-Verzeichniss der Thiere und Vögel, welche von Deppe und Schiede in Mexico gesammelt worden sind.*
Licht. Nordm. Erm. Reis. All.—*Atlas zu Ermann's Reise um die Welt.*
Licht. Verz. der Doubl.—*Lichtenstein, Verzeichniss der Dubletten des zoologischen Museums der Königl. Universität zu Berlin.*
Linn. Syst. Nat.—*Linnæus's Systema Naturæ.*
Lodd.—*Loddiges.*
Lodd. in Proc. of Comm. of Sci. and Corr. of Zool. Soc.—*Loddiges, in the Proceedings of the Committee of Science and Correspondence of the Zoological Society of London.*
Lodd. MSS.—*Loddiges' Manuscripts.*
Longuem. Rev. Zool.—*Longuemare, in Revue Zoologique.*
Long. et Para. Rev. Zool.—*Longuemare and Parzudaki, in Revue Zoologique.*
Martin.—*W. C. L. Martin, A General History of Humming-Birds.*
Mol. Hist. of Chili.—*Molina, History of Chili.*

Montes de Oca in Proc. Acad. Sci. Philad.—*Montes de Oca, in the Proceedings of the Academy of Sciences of Philadelphia.*
Mos. Carls.—*Museum Carlsonianum.*
Mus. Götzian. Dresden.—*The Gotzian Museum, Dresden.*
Natt. in Mus. Vindob.—*Natterer, in the Vienna Museum.*
Nutt. Man.' Orn.—*Nuttall, Manual of Ornithology.*
Para. Rev. Zool.—*Parzudaki, in Revue Zoologique.*
Parzudaki. List of the Trochilidæ.
Pelzeln, Sitz. Acad. Wien.—*Pelzeln, in Sitzungsberichte der Kaiserlichen Akademie der Wissenschaften.*
Penn. Arct. Zool.—*Pennant, Arctic Zoology.*
Prince Max.—*Prince Maximilian zu Wied, Beiträge zur Naturgeschichte von Brasilien.*
Pr. Max. Trav.—*Prince Maximilian's Travels.*
Prinz Maximilian von Wied, Reise nach Brasilien.
Raii—*Ray, in Willughby's Ornithologia.*
Ramon de la Sagra, Hist. de Cuba.—*Ramon de la Sagra, History of Cuba.*
Reich. Av. Syst. Nat.—*Reichenbach's Avium Systema Naturæ.*
Reich. and Reichenb. Aufz. der Col.—*Reichenbach, Aufzählung der Colibris oder Trochilidern, &c., in Cabanis's Journal für Ornithologie.*
Reichenb. in Mus. Heinean.—*Reichenbach, in Heine's Museum.*
Reichenb. Troch. Enum.—*Reichenbach, Trochilinarum Enumeratio.*
Sagra, Voy. de Cub.—*Ramon de la Sagra, Historia fisica, politica y natural de la Isla de Cuba.*
Sallé, Liste des Oiseaux.
Sallé, Rev. Zool.—*Sallé, in Revue Zoologique.*
Salv. in Ibis.—*Salvin, in The Ibis.*
Salv. in Proc. Zool. Soc.—*Salvin, in the Proceedings of the Zoological Society of London.*
Salv. and Sclat. in Proc. Zool. Soc.—*Salvin and Sclater, in the Proceedings of the Zoological Society of London.*
Saucerotte in Mus. Heinean.—*Saucerotte, in Heine's Museum.*
Sauc. or Saucer. MSS.—*Saucerotte, MSS.*
Schmidt, Vögel.
Schomb. Hist. of Barbadoes.—*Sir Richard Schomburgk, History of Barbadoes.*
Sclat. Proc. Zool. Soc.—*Sclater, in the Proceedings of the Zoological Society of London.*
Sclat. and Salv. in Ibis.—*Sclater and Salvin, in The Ibis.*
Shaw, Gen. Zool.—*Shaw's General Zoology.*
Shaw, Mus. Lev. or Leverianum.—*Shaw, Museum Leverianum.*
Shaw, Nat. Misc.—*Shaw, Naturalists' Miscellany.*
Siebold in Mus. Monac.—*Siebold, in the Munich Museum.*
Sloane, Jam.—*Sir Hans Sloane, History of Jamaica.*
Sonn. Œuvres de Buff.—*Œuvres de Buffon, édition par Sonnini.*
Spalowsk. Vögel.—*Spalowsky, Vögel.*
Spix, Av. Sp. Nov. Bras. } *Spix, Avium species novæ in Itinere per Brasiliam.*
Spix, Av. Bras.
Steph. Cont. Shaw's Gen. Zool.—*Stephens's Continuation of Shaw's General Zoology.*
Swains. in Ann. Phil.—*Swainson, in the Annals of Philosophy.*
Swains. Birds of Brazil.—*Swainson, Birds of Brazil.*
Swains. Class. of Birds.—*Swainson, Classification of Birds, in Lardner's Cabinet Cyclopædia.*
Swains. in Phil. Mag.—*Swainson, in the Philosophical Magazine.*
Swains. Syn. Birds of Mexico, in Phil. Mag.—*Swainson's Synopsis of the Birds of Mexico, in the Philosophical Magazine.*
Swains. Zool. Ill.—*Swainson, Zoological Illustrations.*
Swains. Zool. Journ.—*Swainson, in the Zoological Journal.*
Swains. and Rich. Faun. Bor.-Am.—*Swainson and Richardson, Fauna Boreali-Americana, vol. ii. Birds.*
Thévet, les Singularités de la France Antarctique.

Temm. Man. d'Orn. 2nde edit.—*Temminck's Manual d'Ornithologie, second edition.*
Temm. in Mus. Leyden.—*Temminck, in the Leyden Museum.*
Temm. Pl. Col.—*Temminck, Planches Coloriées d'Oiseaux.*
Tschudi, Consp. Av.—*Tschudi, Conspectus Avium.*
Tschudi, Faun. Per.—*Tschudi, Fauna Peruana.*
Turt. edit.—*Turton's edition of Linnæus's Systema Naturæ.*
Valenc. Dict. Sci. Nat.—*Valenciennes, in Le Dictionnaire des Sciences Naturelles.*
Verr. MS.—*Ferreaux's Manuscripts.*
Vieill. Dict. Sci. Nat.—*Vieillot, in Le Dictionnaire des Sciences Naturelles.*
Vieill. Ency. Méth. Orn.—*Vieillot, Tableau Encyclopédie Méthodique, Part III. Ornithologie.*
Vieill. Gal. des Ois.—*Vieillot, Galerie des Oiseaux.*
Vieill. Nouv. Dict. d'Hist. Nat.—*Vieillot, in Le Nouveau Dictionnaire d'Histoire Naturelle.*
Vieill. Ois. de l'Am. Sept.—*Vieillot, Oiseaux de l'Amérique Septentrionale.*
Vieill. Ois. chant. des Amér.—*Vieillot, Histoire Naturelle des Oiseaux chanteurs de la zone torride.*
Vieill. Ois. dor.—*Vieillot, Oiseaux dorés ou à reflets métalliques.*
Vieill et Bonn.—*Bonnaterre and Vieillot, in Tableau Encyclopédie Méthodique, Part III. Ornithologie.*
Vig. in Zool. Journ.—*Vigors, in the Zoological Journal.*
Voy. de la Vénus.—*Voyage de la Vénus.*
Willughby.—*Willughby, Ornithologia.*
Wils. Am. Orn.—*Wilson, American Ornithology.*
Zool. of Beagle.—*The Zoology of the Voyage of H.M.S. Beagle, Part III. Birds, by John Gould.*

LIST

OF

GENERIC AND SPECIFIC NAMES ADOPTED.

	Page
Grypus nævius	35
—— Spixi	35
Eutoxeres Aquila	36
—— Condaminei	37
Glaucis hirsuta	38
—— Mazeppa	38
—— affinis	38
—— lanceolata	39
—— melanura	39
—— Dohrni	39
—— Hueckeri	39
—— Fraseri	39
Threnetes leucurus	40
—— cervinicauda	40
—— Antoniæ	40
Phaëthornis Eurynome	41
—— malaris	41
—— consobrinus	42
—— fraterculus	42
—— longirostris	42
—— syrmatophorus	43
—— Boliviana	42
—— Philippi	43
—— hispidus	43
—— Oseryi	43
—— anthophilus	43
—— Bourcieri	43
—— Guyi	44
—— Emiliæ	44
—— Yaruqui	44
—— superciliosus	45
—— Augusti	45
—— squalidus	45
Pygmornis Longuemareus	46
—— Amaura	46
—— Aspasiæ	47
—— zonura	47
Pygmornis Adolphi	47
—— griseogularis	47
—— strigularis	48
—— Idaliæ	48
—— nigricinctus	48
—— Episcopus	48
—— rufiventris	48
—— Eremita	49
—— pygmæa	49
Eupetomena macroura	50
Sphenoproctus Pampa	51
—— curvipennis	51
Campylopterus lazulus	51
—— hemileucurus	52
—— ensipennis	53
—— splendens	53
—— Villavicencio	53
—— latipennis	53
—— Æquatorialis	54
—— obscurus	54
—— rufus	54
—— hyperythrus	54
Phæochroa Cuvieri	55
—— Roberti	55
Aphantochroa cirrhochloris	55
—— gularis	55
Doleriscæ fallax	56
—— cervina	56
Urochroa Bougueri	56
Sternoclyta cyaneipectus	57
Eugenes fulgens	58
Calligena Clemenciæ	59
Lampoherma Rhami	59
Delattria Henrici	60
—— viridipallens	60
Heliopædica melanotis	60
—— Xantusi	61

	Page		Page
Topaza Pella	61	Microchera albocoronata	82
—— Pyra	62	Lophornis ornatus	82
Oreotrochilus Chimborazo	62	—— Gouldi	83
—— Pichincha	62	—— magnificus	83
—— Estellæ	63	—— Regulus	83
—— leucopleurus	63	—— lophotes	83
—— melanogaster	64	—— Delattrei	84
—— Adelæ	64	—— Reginæ	84
Lampornis Mango	64	—— Helenæ	84
—— iridescens	65	Polemistria chalybea	85
—— Prevosti	65	—— Verreauxi	85
—— Veraguensis	65	Discura longicauda	85
—— gramineus	65	Prymnacantha Popelairei	86
—— viridis	66	Gouldia Langsdorfii	86
—— aurulentus	66	—— Conversi	86
—— virginalis	66	—— Lætitiæ	86
—— porphyrurus	67	Trochilus Colubris	86
Eulampis jugularis	67	—— Alexandri	87
—— holosericeus	68	Mellisuga minima	87
—— chlorolæmus	68	Calypto Costæ	88
—— longirostris	69	—— Annæ	88
Lafresnaya flavicaudata	69	—— Helenæ	88
—— Gayi	69	Selasphorus rufus	88
—— Saulæ	70	—— scintilla	89
Doryfera Johannæ	71	—— Floresii	89
—— Ludovicia	71	—— platycercus	89
—— rectirostris	71	Atthis Heloisæ	89
Chalybura Buffoni	72	Stellula Calliope	90
—— urochrysa	72	Calothorax cyanopogon	90
—— cæruleogaster	72	—— pulchra	91
—— ? Isaurus	72	Acestrura Mulsanti	91
Iolæma frontalis	73	—— decorata	91
—— Schreiberui	73	—— Heliodori	92
Heliodoxa jacula	74	—— micrura	92
—— Jamesoni	74	Chætocercus Rosæ	92
Leadbeatera Otero	74	—— Jourdani	92
—— splendens	74	Myrtis Fanniæ	93
—— grata	75	—— Yarrelli	93
Aithurus Polytmus	75	Thaumastura Coræ	93
Thalurania glaucopis	76	Rhodopis vespera	94
—— Watertoni	76	Doricha Elizæ	94
—— furcata	77	—— Evelynæ	95
—— furcatoides	77	—— enicura	95
—— forficata	77	Tryphæna Duponti	97
—— refulgens	77	Calliphlox amethystina	97
—— Tschudii	78	—— amethystoides	98
—— nigrofasciata	78	—— ? Mitchelli	98
—— venusta	78	Loddigesia mirabilis	99
—— Columbica	78	Spathura Underwoodi	99
—— verticeps	78	—— melananthera	100
—— Fanniæ	78	—— Peruana	100
—— Eriphyle	79	—— rufocaligata	100
—— ? Wagleri	79	—— cissiura	100
Panoplites Jardinei	80	Lesbia Gouldi	101
—— flavescens	80	—— gracilis	101
—— Mathewsi	80	—— Nuna	101
Florisuga mellivora	80	—— Amaryllis	101
—— flabellifera	81	—— eucharis	102
—— atra	81	Cynanthus cyanurus	102

	Page		Page
Cyanthus coelestis	102	Petasophora thalassina	125
—— Mocoa	103	—— cyanotus	125
Cometes sparganurus	103	—— Delphinæ	125
—— Phaon	104	Polytmus viresceus	126
—— ? Glyceria	104	—— viridissimus	127
—— ? Caroli	104	Patagona gigas	127
Pterophanes Temminckii	105	Docimastes ensiferus	129
Aglæactis cupripennis	106	Eugenia Imperatrix	130
—— Æquatorialis	106	Helianthea typica	130
—— parvula	106	—— Bonapartei	130
—— caumatonota	106	—— Eos	131
—— Castelnaudi	107	—— Lutetiæ	131
—— Pamela	107	—— violifera	131
Oxypogon Guerini	108	Heliotrypha Parzudaki	131
—— Lindeni	108	—— viola	131
Ramphomicron heteropogon	109	Heliangelus Clarissæ	132
—— Stanleyi	109	—— strophianus	132
—— Vulcani	109	—— Spencei	132
—— Herrani	109	—— amethysticollis	133
—— ruficeps	109	—— Mavors	133
—— microrhynchus	109	Diphlogæna Iris	133
Urosticte Benjamini	110	—— Aurora	134
Metallura cupreicauda	111	Clytolæma rubinea	134
—— eneicauda	111	—— ? aurescens	134
—— Williami	112	Boureieria torquata	135
—— Primolii	112	—— fulgidigula	135
—— tyrianthina	112	—— insectivora	135
—— Quitensis	112	—— Conradi	135
—— smaragdinicollis	112	—— Inca	136
Adelomyia inornata	113	Lampropygia coeligena	136
—— melanogenys	113	—— Boliviana	137
—— maculata	113	—— purpurea	137
Avocettinus eurypterus	114	—— Prunellei	137
Avocettula recurrirostris	114	—— Wilsoni	137
Anthocephala floriceps	115	Heliomaster longirostris	138
—— ? castaneiventris	115	—— Stuartæ	138
Chrysolampis moschitus	115	—— Sclateri	139
Orthorhynchus cristatus	116	—— pallidiceps	139
—— ornatus	117	—— Constanti	140
—— exilis	117	—— Leocadiæ	140
Cephalepis Delalandi	118	Lepidolarynx mesoleucus	140
—— Loddigesi	118	Calliperidis Angelæ	141
Klais Guimeti	119	Oreopyra leucaspis	141
Myiabeillia typica	119	Eustephanus galeritus	141
Heliactin cornuta	120	—— Stokesi	142
Heliothrix auritus	121	—— Fernandensis	142
—— auriculatus	121	Phæolæma rubinoides	142
—— phainolæma	121	—— Æquatorialis	143
—— Barroti	121	Eriocnemis cupreiventris	143
—— violifrons	122	—— Isaacsoni	144
Schistes Geoffroyi	122	—— Luciani	144
—— personatus	122	—— Mosquera	144
—— albigularis	123	—— vestita	145
Augastes scutatus	123	—— nigrivestis	145
—— Lumachellus	123	—— Godini	145
Petasophora serrirostris	124	—— D'Orbignyi	145
—— Anais	124	—— Derbiana	145
—— iolata	124	—— Alinæ	145
—— coruscans	125	—— squamata	146

o 2

	Page		Page
Eriocnemis lugens	145	Chrysuronia Nvera	165
— Aurelia	146	— cæruleicapilla	165
Cyanomyia quadricolor	147	— Humboldti	165
— violiceps	147	— Elicia	165
— cyanocephala	147	— chrysura	165
— Guatemalensis	148	Eucephala Grayi	166
— Francia	149	— smaragdo-cærulea	166
— cyanicollis	149	— chlorocephala	166
Hemistilbon Ocai	150	— cæruleo-lavata	166
— Norrisi	150	— scapulata	166
Leucippus chionogaster	150	— hypocyanea	166
Leucochloris albicollis	151	— cærulea	167
Thaumatias candidus	151	— cyanogenys	167
— chionopectus	152	Panterpe insignis	168
— leucogaster	152	Juliamyia typica	168
— viridiceps	152	— Feliciana	168
— Milleri	152	Circe latirostris	169
— nitidifrons	152	— Doubledayi	169
— cæruleiceps	152	Phæoptila sordida	169
— breviroetris	152	— zonura	170
— affinis	153	Damophila amabilis	170
— chionurus	153	Hylocharis sapphirina	171
— albiventris	153	— lactea	171
— Linnæi	153	— cyanea	171
— fluviatilis	154	Sapphironia Goudoti	172
— apicalis	154	— cæruleigularis	172
— maculicaudus	154	Sporadinus elegans	173
Amazilia pristina	155	— Ricordi	173
— alticola	156	— ? Maugæi	173
— Dumerili	156	Chlorolampas auriceps	174
— leucophæa	156	— Caniveti	174
Pyrrhophæna cinnamomea	156	— Osberti	174
— Yucatanensis	157	— Salvini	174
— cerviniventris	157	Chlorostilbon angustipennis	175
— castaneiventris	157	— Hæberlini	175
— Rieffori	158	— Phaethon	175
— beryllina	158	— aureiventris	176
— Devillei	158	— prasinus	176
— viridigaster	159	— igneus	176
— iodura	159	— Atala	177
— cyanura	160	— Daphne	177
Erythronota antiqua	160	— Peruanus	177
— Feliciæ	161	— Naponsis	177
— Edwardi	161	— breviscudatus	178
— niveiventris	161	— chrysogaster	178
— elegans	162	— assimilis	178
Saucerottia typica	162	— nitens	179
— Sophia	162	Panychlora Aliciæ	179
— Warszewiczi	163	— euchloris	179
Hemithylaca cyanifrons	163	— stenura	180
Eupherusa eximia	163	— Poortmanni	180
Chrysuronia Œnone	164	Smaragdochrysis iridescens	181
— Josephinæ	164	Phlogophilus hemileucourus	181

INDEX OF GENERIC NAMES ADOPTED.

Name	Page	Name	Page	Name	Page
Acestrura	41	Eugenes	57	Oreopyra	141
Adelomyia	113	Eugenia	129	Oreotrochilus	62
Aglæactis	105	Eulampis	67	Orthorhynchus	116
Aithurus	76	Eupetomena	69	Oxypogon	107
Amazilis	155	Eupherusa	153	Panoplites	70
Anthocephala	114	Eustephanus	141	Panterpe	107
Aphantochroa	85	Eutoxeres	36	Panychlora	179
Anthis	89	Florisuga	80	Patagona	127
Augastes	123	Glaucis	27	Petasophora	124
Avocettinus	114	Gouldia	86	Phæochroa	84
Avocettula	114	Grypus	35	Phæolæma	142
Bourcieria	135	Heliactin	120	Phæoptila	160
Calliperidia	140	Heliangelus	132	Phaëthornis	41
Calliphlox	97	Helianthea	130	Phlogophilus	181
Calothorax	90	Heliodoxa	74	Polemistria	84
Calypte	67	Heliomaster	138	Polytmus	126
Campylopterus	81	Heliopædica	60	Prymnacantha	86
Cephalepis	118	Heliothrix	120	Pterophanes	105
Chætocercus	92	Heliotrypha	131	Pygmornis	46
Chalybura	72	Hemistilbon	149	Pyrrhophæna	158
Chlorolampis	273	Hemithylaca	163	Ramphomicron	109
Chlorostilbon	175	Hylocharis	171	Rhodopis	94
Chrysolampis	115	Iolæma	73	Sapphironia	179
Chrysuronia	164	Juliamyia	168	Saucerottia	162
Circe	108	Klais	119	Schistes	122
Clytolæma	134	Lafresnaya	69	Sciasphorus	98
Cœligena	59	Lampornis	64	Smaragdochrysis	160
Cometes	103	Lamprolæma	59	Spathura	99
Cyanomyia	147	Lampropygia	136	Sphenoproctus	50
Cyanthus	102	Leadbeatera	74	Sporadinus	172
Damophila	170	Lepidolarynx	140	Stellula	90
Delattria	59	Leshia	101	Sternoclyta	57
Diphlogæna	133	Leucippus	150	Thalurania	78
Discura	85	Leucochloris	150	Thaumastura	93
Docimastes	129	Loddigesia	99	Thaumatias	151
Doleriaca	56	Lophornis	82	Threnetes	40
Doricha	94	Mellisuga	87	Topaza	61
Doryfera	71	Metallura	111	Trochilus	66
Eriocnemis	143	Microchera	82	Tryphæna	98
Erythronota	160	Myiabeillia	119	Urochroa	58
Eucephala	166	Myrtis	92	Uroticto	110

INDEX

OF

SPECIFIC NAMES

OF

HUMMING-BIRDS.

[The following is an alphabetical arrangement of all the specific names with which I am acquainted; those appearing in my own work, (and to which the numbers refer,) and others which I have not been able to determine to what species they have been applied. Some of the latter have been assigned to manufactured specimens, and the descriptions of the others are frequently so curt and vague that they cannot be identified.]

		PAGE
Abrillei	Myiabeillia typica	119
abnormis, *Natt.*	Not identified.	
acuticaudus	Panychlora acuura	130
Addæ	Spathura rufocaligata	100
Adela	} Oreotrochilus Adelæ	64
Adelæ		
Adolphi	Pygmornis Adolphi	47
æneicauda	} Metallura æneicauda	111
æneicaudus		
æneocauda		
Æquatorialis	Agleactis Æquatorialis	106
——	Campylopterus Æquatorialis	54
——	Pheolæma Æquatorialis	143
affinis	Thaumatias affinis	163
——	Glaucis affinis	38
——	Phaëthornis superciliosus	46
Aglaiæ	Pyrrhophæna iodura	150
albicollis	Leucochloris albicollis	151
albigularis	Leucochloris albicollis	151
——	Schistes albigularis	143
albirostris	Thaumatias leucogaster	162
albiventris	Thaumatias albiventris	153
albocoronata	Microchera albocoronata	83
albus	Lampornis Mango	64
Alexandri	Trochilus Alexandri	87
Aliciæ	} Panychlora Aliciæ	179
Aliciæ		

		PAGE
Alinæ	} Eriocnemis Alinæ	145
Alinæ . . . . .
Aline . . . .
Allardi	Metallura tyrianthina	112
alticola	Amazilia alticola	156
amabilis	Damophila amabilis	170
Amaryllis	Lesbia Amaryllis	101
amaura	Pygmornis amaura	46
amazicula	Amazilia Dumerili	156
Amazili	Amazilia pristina	155
—	Pyrrhophæna Riefferi	158
Amazilis	Amazilia pristina	156
amazilicula . . .	Amazilia Dumerili	156
amethysticollis . .	Heliangelus amethysticollis . . .	113
amethystina . .	} Calliphlox amethystina	97
amethystinus . .
amethystoides . .	Calliphlox amethystoides	98
Anais	Petasophora Anais	124
—	Petasophora cyanotis	125
—	Petasophora iolata	124
—	Petasophora thalassina	125
Angela	} Calliperidia Angelæ	141
Angelæ . . . .
angustipennis . .	Chlorostilbon angustipennis . . .	175
Anna	} Calypte Annæ	88
Annæ . . . .
anthophila . .	} Phaëthornis anthophilus	43
anthophilus . .
antiqua	Erythronota antiqua	160
Antoniæ	Threnetes Antoniæ	40
apicalis	Phaëthornis Guyi	44
—	Thaumatias apicalis	154
Aquila	Rutoxeres Aquila	36
Arsennii	Heliopædica melanotis	60
Aranoë	Pyrrhophæna beryllina	158
araneoides . . .	Pyrrhophæna Riefferi	158
Aspasia	Pygmornis Aspasiæ	47
assimilis	Chlorostilbon assimilis	178
Atala	Chlorostilbon Atala	177
—	Chlorostilbon chrysogaster . . .	178
Atalæ	Chlorostilbon Atala	177
ater . . . .
atra	} Florisuga atra	81
atratus . . . .
atricapillus . . .	Lampornis Mango	65
atrigaster . . .	Eulampis holosericeus	68
atrimentalis . . .	Pygmornis Amaura	46
Audebertii . . .	Eucephala cærulea	167
Audeneti	Polemistria chalybæa	83
Augusta . . . .
Augustæ	} Phaëthornis Augusti	45
Augusti . . . .
aurantius, Gmel. . .	Not determined.
aurata	Panychlora Aliciæ	179
auratus	Eulampis jugularis	67
—	Lophornis ornatus	82
aureiventris . . .	Chlorostilbon aureiventris	176
Aureliæ	Eriocnemis Aureliæ	146
aureoviridis . . .	Lampornis viridis.
aurescens . . .	Clytolæma ? aurescens	134
aureus, Licht. . . .	Clytolæma rubinea ?.

		PAGE
auriceps	Chlorolampis auriceps	174
auriculata	} Heliothrix auriculatus	121
auriculatus		
aurigaster	Eulampis holosericeus	68
——	Heliantheа Bonapartei	130
aurita	Heliothrix auritus	121
——	Heliothrix auriculatus	121
auritus	Eulampis jugularis	68
——	Heliothrix auritus	121
Aurora	Diphlogena Aurora	134
aurulenta	} Lampornis aurulentus	60
aurulentus		
Avocetta	Avocettula recurvirostris	114
Azaræ, Vieill.	Not determined.	
azureus	Hylocharis cyaneus	171
Bahamensis	Doricha Evelynæ	93
Bancrofti	Eulampis jugularis	67
Barroti	Heliothrix Barroti	121
——	Heliothrix violifrons	121
Benjamini	} Trosticlo Benjamini	110
Benjaminus		
beryllina	} Pyrrhophæna beryllina	158
beryllinus		
bicolor	Hylocharis cyaneus	171
——	Thalurania? Wagleri	70
bifurcata	Lesbia eucharis	102
biloba		
bilopha	} Heliactin cornuta	120
bilophus		
bipartitus, Less.	Cynanthus cyanurus?	
Boliviana	Lampropygia Boliviana	137
——	Phaëthornis Boliviana	42
bombilus	Acestrura Heliodori	97
Bonapartei	Heliantheа Bonapartei	130
Boothi	Calypte Helenæ	88
Bougueri	Urochroa Bougueri	56
Bourcieri	Phaëthornis Bourcieri	43
brachyrhynchus	Ramphomicron microrhynchus	109
Brasiliauns	Glaucis hirsuta	38
Brasiliensis	Pygmornis Eremita	49
——	Phaëthornis squalidus	46
——	Phaëthornis superciliosus	45
——	Glaucis hirsuta	38
brevicaudatus	Chlorostilbon brevicaudatus	178
brevicaudus	Calliphlox amethystina	98
brevirostris	Thaumatias brevirostris	152
bromicolor	Lampornis porphyrurus	67
Buffoni	Chalybura Buffoni	72
cærulea	Eucephala cærulea	107
cæruleicapilla	Chrysuronia cæruleicapilla	103
cæruleiceps	Thaumatias cæruleiceps	152
cæruleigaster	Chalybura cæruleogaster	72
cæruleigularis	Sapphironia cæruleigularis	172
cæruleiventris	} Chalybura cæruleogaster	72
cæruleogaster		
cæruleogularis	Sapphironia cæruleigularis	172
cæruleo-lavata	Eucephala cæruleo-lavata	166
cæruleucens	Sapphironia cæruleigularis	172
cæruleus	Eucephala cærulea	107
caligatus	Saucerottia Sophiæ	162
Calliope	Stellula Calliope	90

		PAGE
campestris	Calliphlox amethystina	48
campylopters	Sphenoproctus Pampa	51
campylopterus	Campylopterus latipennis	53
campylostylus	Aphantochroa cirrhochloris	55
candida }	Thaumatias candidus	151
candidus		
Caniveti	Chlorolampis Caniveti	174
Capensis, Gmel.	Not determined.	
carbunculus	Chrysolampis moschitus	115
Caroli }	Cometes ? Caroli	104
Carolus		
Casini	Trochilus Alexandri	87
castaneiventris	Anthocephala ? castaneiventris	115
——	Phyrrhophaena castaneiventris	157
castaneocauda	Heliopaedica Xantusi	61
castaneuventris	Anthocephala ? castaneiventris	115
Castelnaudi		
Castelnaul }	Aglaactis Castelnaudi	107
Castelnaui		
Catharinae	Melliauga minima	87
caudacutus, Vieill.	Not determined.	
caumaionots }	Aglaactis caumaionots	100
caumaionotus		
Cecilia	Oreotrochilus Estelle	83
cephalatra	Aithurus polytmus	75
cephalus	Phaëthornis longirostris	42
cervina	Doleriana cervina	94
cervinicauda	Threnetes cervinicauda	40
cerviniventris	Pyrrhophaena cerviniventris	157
chalcotis	Petasophora serrirostris	123
chalybea }	Polemistria chalybea	85
chalybeus		
Chimboraso	Oreotrochilus Chimboraso	102
chionogaster	Leucippus chionogaster	130
chionopectus	Thaumatias chionopectus	152
chionura }	Thaumatias chionurus	155
chionurus		
chlorocephala }	Eucephala chlorocephala	106
chlorocephalus		
chlorolaemus }	Eulampis chlorolaemus	68
chlorolaemus		
chlorolaimus		
chlorolucurus	Polytmus virescens	127
chlorolophus	Orthorhynchus exilis	117
chloropogon	Urolampra chloropogon, Cab. et Hein.	
	Not seen.	
chrysobronchus	Polytmus virescens	120
chrysochloris	Cometes sparganurus	104
chrysogaster	Chlorostilbon chrysogaster	174
——	Chlorostilbon Haeberlini	175
——	Helianthea Bonapartei	140
chrysogastra	Chlorostilbon chrysogaster	174
chrysolopha	Heliactin cornuta	120
chrysura }	Chrysuronia chrysura	165
chrysurus		
——	Grypus Spixi	35
——	Cometes sparganurus	103
——	Polytmus viridissimus	127
ciliatus, Lath. MS.	Not determined.	
cinereicollis, Vieill.	Not determined.	
cinereus	Campylopterus latipennis	53

		PAGE
cinnamomeus	Eustephanus Fernandensis	142
——	Pyrrhophæna cinnamomea	156
Circe	Circe latirostris	103
cirrhochloris }	Aphantochroa cirrhochloris	55
cirrochloris		
cisiura	Spathura cisiura	100
Clarissæ		
Clarissæ }	Heliangelus Clarissæ	132
Clarissæ		
Clemenciæ	Cœligena Clemenciæ	50
Cleopatra	Thaumatias lenongaster	152
cœlestis	Cyanthus cœlestis	103
——	Tryphæna Duponti	97
cœligena	Lampropygia cœligena	136
cœlina	Sapphironia cœrulsigularis	172
cœrulcirentris	Chalybura cœruleogaster	72
collaris	Selasphorus rufus	88
colubris	Trochilus colubris	86
Colombica		
Columbiana }	Thalurania Columbica	78
Columbicus		
concinna, Steph.	Mellisugus, " Lies." Reichenbach.	
Condaminei }	Entoxeres Condaminei	97
Condamini		
Conradi	Bourcieria Conradi	136
consobrinus	Phaëthornis consobrinus	42
Constantii	Heliomaster Constantii	140
Conversii	Gouldia Conversii	80
conurus, Steph.	Not determined.	
Coræ }	Thaumastura Coræ	93
Coræ		
corallirostris	Pyrrhophæna cinnamomea	156
Corinna, Less.	Heliomaster longirostris	
cornuta }	Heliactin cornuta	120
cornutus		
coruscans	Petasophora coruscans	125
coruscus	Ramphomicron heteropogon	109
——	Calothorax cyanopogon	91
Costæ	Calypte Costæ	88
crispa }	Petasophora serrirostris	124
crispus		
cristata	Orthorhynchus cristatus	117
cristatellus	Orthorhynchus exilis	117
cristatus	Orthorhynchus cristatus	116
cryptarus	Panychlora Aliciæ	170
cuculliger	Heliopædica melanotis	60
cupreicauda }	Metallura cupreicauda	111
cupreicaudus		
cupreiventris	Eriocnemis copraiventris	143
cupreocauda	Metallura cupreicauda	111
cupreoventris	Eriocnemis cupreiventris	144
cuprienda	Metallura cupreicauda	111
cupripennis	Aglæactis cupripennis	106
cupreiventris	Eriocnemis cupreiventris	141
curvipennis	Sphenoproctus curvipennis	51
Cuvieri	Phæochroa Cuvieri	55
cyanea	Hylocharis cyanea	171
cyanæipectus		
cyanipectus }	Sternoclyta cyanipectus	57
cyanopectus		
cyaneus	Eulampis jugularis	68

		PAGE
cyaneus	Hylocharis cyanea	171
cyanicollis	Cyanomyia cyanicollis	149
cyanifrons	Hemithylaca cyanifrons	153
cyanocephala	Cyanomyia cyanocephala	147
——	Cyanomyia Guatemalensis	144
——	Cyanomyia quadricolor	147
——, Gmel.	Not determined.	
cyanocollis	Cyanomyia cyanicollis	149
cyanogenys	Eucephala cyanogenys	167
cyanomelas	Eulampis jugularis	67
	Sapphironia cœruleigularis	172
cyanopectus	Sternoclyta cyaneipectus	5
cyanopogon	Calothorax cyanopogon	00
cyanopterus	Pterophanes Temmincki	103
cyanopygus	Eriocnemis cupreiventris, Reichenb.	
cyanotis	} Petasophora cyanotis	125
cyanotus		
cyanura	Pyrrhophæna cyanura	160
cyanurus	Cynanthus cyanurus	102
——, Gmel.	Not determined.	
cyanus, Vieill.	Not determined.	
Daphne	Chlorostilbon Daphne	177
dasypus	Eriocnemis Alinæ	146
Davidianus	Pygmornis rufiventris	48
decorata	} Acestrura decorata	91
decoratus		
decorus	Lophornis magnificus	83
De Filippi	Phaëthornis Philippi	43
Delalandi	Cephalepis Delalandi	118
Delattre	} Campylopterus hemileucurus	52
Delattrei		
——	Lophornis Delattrei	84
Delphina	} Petasophora Delphinæ	125
Delphinæ		
Derbianus	Docimastes ensiferus	120
——	Eriocnemis Derbiana	145
Derbyanus	Docimastes ensiferus	120
——	Eriocnemis Derbianus	145
Derbyi	Eriocnemis Derbiana	145
Devillei	Pyrrhophæna Devillei	158
dichrous, Licht.	Chrysuronia chrysura, Reichenb.	
dilophus	Heliactin cornuta	120
dispar	Heliotrypha Parzudaki	111
Dohrni	Glaucis Dohrni	39
Dominica	Lampornis gramineus	66
Dominicensis	Mellisuga minima	87
Dominicus	Lampornis aurulentus	66
——	Glaucis hirsuta	38
——	Lampornis gramineus	66
D'Orbignyi	Eriocnemis D'Orbignyi	145
Doubledayi	Circe Doubledayi	160
——	Circe latirostris	160
Dubusi	Pyrrhophæna Riefferi	158
Duchassaigni	Sapphironia cœruleogularis	172
Dumerili	Amazilia Dumerili	141
——	Pyrrhophæna Devillei	150
Duponti	Tryphæna Duponti	97
Edwardi	Erythronota Edwardi	161
elatus	Chrysolampis moschitus	115
elegans	Sporadinus elegans	173
——	Erythronota elegans	162

		PAGE
Elicia	} Chrysuronia Elicia	103
Elicis		
Elisa	} Doricha Elizæ	24
Elizæ		
Emilæ	Erythronota Felicia	161
Emiliæ	Phaëthornis Emiliæ	44
enicura	} Doricha enicura	25
enicurus		
ensifera	} Docimastes ensiferus	129
ensiferus		
ensipennis	Campylopterus ensipennis	53
Eos	Helianthea Eos	131
Episcopus	Pygmornis Episcopus	48
Eremita	Pygmornis Eremita	49
Eriphyle		
Eryphila	} Thalurania Eriphyle	79
Eryphile		
erythronota		
erythronotos	} Erythronota antiqua	160
erythronotus		
erythrorhyncha, Bp.	Not a species.	
Esmeralda	Panychlora Poortmanni	180
Estella	Oreotrochilus Estellæ	63
Estellæ		
euanthes	Ramphomicron microrhynchus	110
eucharis	Lesbia eucharis	102
euchloris	Panychlora euchloris	180
Eurynome	} Phaëthornis Eurynome	41
Eurynomus		
eurypterus	} Avocettinus eurypterus	114
earypterus		
Evelinæ		
Evellinæ	} Doricha Evelynæ	25
Evelynæ		
excisus	Thalurania Eriphyle	79
exilis	Orthorhynchus exilis	117
eximia	} Eupherusa eximia	163
eximius		
exortis	Heliotrypha Parzudaki	131
falcata	} Campylopterus lazulus	51
falcatus		
fallax	Dolerisca fallax	50
Fanniæ	} Thalurania Fanniæ	79
Fannyi		
Fanny	Myrtis Fanniæ	93
fasciatus	Lampornis Mango	65
——, Shaw	Not determined.	
Faustinæ	Cyanomyia cyanocephala	147
Felicia	} Erythronota Felicia	161
Feliciæ		
Feliciana	Juliamyia Feliciana	168
Fernandensis	Eustephanus Fernandensis	142
ferrugineus	Glaucis hirsuta	36
festivus	Polemistria chalybea	85
filicaudus	Acestrura Mulsanti	91
fimbriata	} Florisuga mellivora	80
fimbriatus		
flabellifera	} Florisuga flabellifera	81
flabelliferus		
flammifrons	Rostephanus galeritus	142
flavescens	Panoplites flavescens	80

		PAGE
flavicaudata . . . . }	Lafresnaya flavicaudata	60
flavicaudatus . . .		
flavifrons	Chlorostilbon Phaëthon	175
——, Gmel. . . .	Not determined.	
Floresii	Lampornis porphyrurus	67
——	Selasphorus Floresii	89
floriceps	Anthocephala floriceps	115
fluviatilis	Thaumatias fluviatilis	164
forcipatus	Eupetomena macroura	60
forficata	Thalurania forficata	77
forficatus	Eustephanus galeritus	141
——	Cynanthus cyanurus	102
Franciæ	Cyanomyia Franciæ	140
Fraseri	Glaucis Fraseri	89
fraterculus . . .	Phaëthornis fraterculus	42
frontalis	Iolæma frontalis	73
——	Thalurania glaucopis	70
fulgens	Eugenes fulgens	58
fulgidigula	Bourcieria fulgidigula	115
fulgidus	Lampraleme Rhami	64
fulvifrons	Hylocharis sapphirina	171
fulviventris	Doleriscs fallax	66
fulvus, Gmel. . . .	Not determined.	
furcata	Thalurania furcata	77
furcatoides . . .	Thalurania furcatoides	77
furcatus	Thalurania furcata	77
——	Thalurania Tschudii	78
furcifer, Shaw . .	Not determined.	
fusca	Florisuga atra	81
fuscicaudatus . . .	Pyrrhophæna Riefferi	158
fuscus	Florisuga atra	81
Gabriel	Heliothrix Barroti	122
Galathea	Chlorostilbon prasinus	170
galerita . . . . }	Eustephanus galeritus	141
galeritus		
Gayi	Lafresnaya Gayi	60
Geoffroyi	Schistes personatus	124
——	Schistes Geoffroyi	123
Georgina . . . . }	Avocettinus eurypterus	114
Georginæ		
Gibsoni, Fras. . .	A manufactured specimen.	
gigantea . . . . }	Patagona gigas	128
gigas		
glaucopis	Thalurania glaucopis	70
glaucopoides, D'Orb. et Lafres. . . . }	Not determined.	
glomata	Eriocnemis vestita	145
Glyceria	Cometes ? Glyceria	104
Godini	Eriocnemis Godini	145
Gorgo	Cynanthus cyanurus	102
Goudoti	Sapphironia Goudoti	172
Gouldi	Lesbia Gouldi	101
——	Lophornis Gouldi	83
——	Petasophora serrirostris	124
gracilis	Lesbia gracilis	101
gramineus . . . . }	Lampornis gramineus	65
granadensis . . .	Phæolæma rubinoides.	
granatinus	Eulampis jugularis	67
grata	Leadbeatera grata	75
Grayi	Eucephala Grayi	106

203

		PAGE
griseigularis	} Pygmornis griseogularis	47
griseogularis		
Guatemalensis	Cyanomyia Guatemalensis	148
Guerini	Oxypogon Guerini	106
Guimeti	Klais Guimeti	110
Gujanensis	Chrysolampis moschitus	115
gularis	Aphantochroa gularis	50
——	Lampornis gramineus	06
gutturalis	Lampornis Mango ?.	
Guy	} Phaëthornis Guyi	44
Guyi		
Gyrinus	Thalurania furcata	77
Haeberlinii	Chlorostilbon Haeberlini	175
hæmatorhynchus, Bp.	Not a species.	
Helena	Lophornis Helenæ	84
Helenæ	Calypte Helenæ	88
——	Lophornis Helenæ	84
Helianthea	Helianthea typica	130
Heliodori	Acestrura Heliodori	12
helios	Lophornis magnificus	83
Heloisa	} Atthis Heloisæ	80
Heloisæ		
hemileucurus	Campylopterus hemileucurus	59
——	Phlogophilus hemileucurus	181
Henricæ	} Delattria Henrici	60
Henrici		
Herrani	Ramphomicron Herrani	100
heteropogon	Ramphomicron heteropogon	100
heteropygia	Doricha enicura	96
hirsuta	} Glaucis hirsuta	38
hirsutus		
hirundinaceus	Eupetomena macroura	50
hirundinaceus	Gouldia Langsdorffi	80
hispida	} Phaëthornis hispidus	43
hispidus		
Hoffmanni	Saucerottia Sophiæ	162
holosericeus	} Eulampis holosericeus	08
holosericeus		
Humboldti	Chrysuronia Humboldti	155
humilis	Mellisuga minima	87
hyperythrus	Campylopterus hyperythrus	64
hypocyaneus	Eucephala hypocyanea	166
hypoleucus	Leucippus chionogaster	150
——	Cyanomyia Franciæ	149
hypophæus	Chrysolampis moschitus	115
kteroceephalus	Calypte Annæ	83
Idaliæ	Pygmornis Idaliæ	48
igneus	Chlorostilbon igneus	170
Imperatrix	Eugenia Imperatrix	130
Inca	Bourcieria Inca	130
inornata	} Adelomyia inornata	113
inornatus		
insectivora	} Bourcieria insectivora	135
insectivorus		
insignis	Panterpe insignis	168
intermedius	} Phaëthornis squalidus	45
intermedius		
iodura	} Pyrrhophæna iodura	150
iodurus		
iolata	} Petasophora iolata	124
iolatus		

		PAGE
iridescens	Lampornis iridescens	65
	Smaragdochrysis iridescens	161
Iris	Diphlogena Iris	153
Isaacsoni	Eriocnemis Isaacsoni	144
Isaure	Chalybura? Isaure	72
jacula	Heliodoxa jacula	74
Jamesoni	Heliodoxa Jamesoni	74
—	Oreotrochilus Pichincha	63
janthinotus	Petasophora serrirostris	124
Jardinei } Jardini }	Panoplites Jardinei	80
Johannæ	Doryfera Johannæ	71
iolata	Petasophora iolata	124
Josephinæ	Chrysuronia Josephinæ	164
Jourdani	Chætocercus Jourdani	22
jugularis	Eulampis jugularis	67
Julia Juliæ Julis }	Juliamyia typica	166
Kienerii	Spathura Underwoodi	100
Kingii	Cynanthus cyanurus	102
—	Eustephanus galeritus	141
Labrador	Myrtis Fanny	83
lactea	Hylocharis lactea	171
—	Thaumatias Linnæi	154
Lætitia Lætitiæ }	Gouldia Lætitiæ	86
Lafresnayi	Lafresnaya flavicaudata	69
Lalandi	Cephalepis Delalandi	113
lamprosephalus	Calypte Annæ	88
lamprogenesis, Bonap.	Lampornis Prevosti, Reichenb.	
lampros	Chlorostilbon præsinus	178
lanceolatus	Glaucis lanceolatus	32
Langsdorffi	Gouldia Langsdorffi	86
largipennis	Campylopterus latipennis	53
lasiopygus	Heliotrypha Parzudaki	131
latipennis	Campylopterus latipennis	53
—	Campylopterus ensipennis	53
latirostris	Amazilis pristina	156
	Hylocharis sapphirina	171
	Circe latirostris	170
lanula	Circe Doubledayi	169
	Circe latirostris	169
laulissus	Hylocharis lactea	171
lanulas	Campylopterus lanulus	61
	Circe Doubledayi	171
—	Circe latirostris	169
	Lampornis Mango	65
Leadbeateri	Leadbeatera grata	75
Leocadiæ	Heliomaster Leocadiæ	140
lepida lepidus }	Trypheena Duponti	97
Lereboulleti	Circe Doubledayi	169
Lessoni	Avocettula recurvirostris	114
	Cyanomyia cyanocephala	147
	Circe latirostris	169
leucaspis	Oreopyra leucaspis	141
leuocrotaphus	Heliopædica melanotis	60
	Heliothrix auritus	121
leucogaster	Leucippus chionogaster	150
—	Thaumatias leucogaster	152

		PAGE
leucogastra	Thaumatias leucogaster	153
leucophæa	Amazilia leucophæa	150
leucophrys	Phaëthornis squalidus	45
leucopleurus	Oreotrochilus leucopleurus	03
leucopterus	Oreotrochilus leucopleurus	04
leucopygius	Florisuga atra	81
leucotis	Heliopædica melanotis	00
leucurus	Threnetes leucurus	40
Libussa	Heliangelus Clarissæ	133
Lichtensteinii	Panoplites flavescens	80
ligonicauda, ligonicaudus	} Discura longicauda	85
Lindeni	Oxypogon Lindeni	108
Linnæi	Thaumatias Linnæi	153
Loddigesii	Cephalepis Loddigesi	118
longicauda, longicaudus	} Discura longicauda	85
longirostris	Eulampis longirostris	09
——	Heliomaster longirostris	138
——	Heliomaster Stuartæ	138
——	Heliomaster mesoleucus	140
——	Phaëthornis longirostris	42
Longuemarei, Longuemareus, Longuemari	} Pygmornis Longuemareus	46
lophotes	Lophornis lophotes	83
Luciani	Eriocnemis Luciani	144
lucida, lucidus	} Heliopædica melanotis	61
——	Cœligena Clemenciæ	59
Lucifer	Calothorax cyanopogon	90
Ludoviciæ	Doryfera Ludoviciæ	71
lugens	Eriocnemis lugens	140
lugubris	Florisuga atra	81
Lumachella, Lumachellus	} Augastes Lumachellus	127
Lutetiæ	Helianthea Lutetiæ	131
Lydia	Thalurania verticeps	78
macroura, macrourus, macrurus	} Eupetomena macroura	50
maculata	Adelomyia maculata	113
——	Thaumatias Linnæi	154
maculatum	Grypus nævius	35
maculatus	Lampornis gramineus	65
——	Thaumatias Linnæi	153
maculicaudus	Thaumatias maculicaudus	154
maculicollis	Panyochlora Aliciæ	170
magnifica, magnificus	} Lophornis magnificus	83
malaris	Phaëthornis malaris	41
Malvina	Not seen.	
Mango	Lampornis Mango	64
——	Lampornis porphyrurus	67
maniculata	Eriocnemis cupreiventris	144
maniculatum	Grypus nævius	35
margaritaceus	Lampornis aurulentus	66
Maria, Mariæ	} Pyrrhophæna Devillei	150
——	Aithurus polytmus	75
marmoratus	Lampornis gramineus	66

		PAGE
Matthewsi	Panoplites Matthewsi	80
Maugeei	} Sporadinus ? Maugei	173
Maugæus	
Maugeana	} Sporadinus ? Maugæi	
Maugeanus	
Maugei	Thaumatias Linnei	154
Mavors	Heliangelus Mavors	133
maxillosus	Phaëthornis malaris ?	
maximus, *Verill.*	Not determined.	
Maynensis	Leadbeatera Otero, *Reichenb.*	74
Mazeppa	Glaucis Mazeppa	38
melananthera	Spathura melananthera	110
melanogaster	Oreotrochilus melanogaster	64
—	Eugenes fulgens	64
melanogenys	Adelomyia melanogenys	113
melanolophus, *Verill.*	Not determined.	
melanorhynchus	Chlorostilbon chrysogaster	178
melanotis	Phaëthornis Eurynome	41
melanotis	} Hellopædica melanotis	60
melanotus	
melanura	Glaucis melanura	39
mellisugus	Chlorostilbon Atala	177
—	Thaumatias leucogaster	152
mellivora	} Florisuga mellivora	80
mellivorus	
Meriphile	Thalurania Eriphyle	70
Merritzii	Klais Guimeti	119
mesoleuca	} Lepidolarynx mesoleucus	140
mesoleucus	
metallicus	Chlorostilbon Phaëthon	175
Mexicanus	Eulampis holosericeus	68
microrhyncha	
microrhynchum	} Ramphomicron microrhynchus	100
microrhynchus	
micrura	} Acestrura micrura	92
micrurus	
Milleri	Oreotrochilus leucopleurus	64
—	Thaumatias Milleri	152
minima	} Mellisuga minima	87
minimus	
minullus, *Verill.*	Not determined.	
mirabilis	Loddigesia mirabilis	99
Mitchelli	Calliphlox ? Mitchelli	98
Mocoa	Cynanthus Mocoa	103
modestus	Chlorolampis auriceps	174
montana	} Selasphorus platycercus	82
montanus	
Moorei	Phaëthornis consobrinus	42
moschita	} Chrysolampis moschitus	115
moschitus	
Mosquera	Eriocnemis Mosquera	144
mosquitus	Chrysolampis moschitus	115
Mossei	Cometes ? Glyceria	104
Mulsanti	Acestrura Mulsanti	91
multicolor, *Gmel.*	Not determined.	
mystacinus	Lepidolarynx mesoleucus	140
myttax	Polemistra chalybæa	85
nævia	} Grypus nævius	35
nævius	
Napensis	Chlorostilbon Napensis	177
Nattereri	Augastes scutatus	123

Nerra	Chrysuronia Nerra	163
niger	Florisuga atra	81
nigra	Mellisuga minima	87
nigra	Mellisuga minima	87
nigricincta }	Pygmornis nigricinctus	48
nigricinctus }		
nigricollis	Lampornis Mango	65
nigrirostris, *Reichenb.*	Not determined.	
nigriventris	Eriocnemis nigriventris	145
nigrofasciata }	Thalurania nigrofasciata	78
nigrofasciatus }		
nigrotis	Heliothrix auritus	121
nitens	Chlorostilbon nitens	172
nitidifrons	Thaumatias nitidifrons	153
nitidissimus	Chlorostilbon prasinus	170
nitidus	Lampornis Mango	65
niveipectus	Thaumatias chionopectus	153
niveiventer	Erythronota niveiventris	162
niveiventris	Erythronota niveiventris	161
niveoventer	Erythronota niveiventris	161
Norrisi	Hemistilbon Norrisi	150
Nuna	Lesbia Nuna	101
obscura	Pygmornis Idaliæ	48
obscurus	Campylopterus obscurus	64
——	Clytolæma rubinea	134
——, *Gmel.*	Not determined.	
Ocai	Hemistilbon Ocai	150
ochropygos, *Natt.*	Phaëthornis Pretrei, *Reichenb.*	
Œnone	Chrysuronia Œnone	164
opaca	Metallura cupricauda	111
opisthococcus	Cephalepis Loddigesi	118
Ortignyi	Eriocnemis D'Orbignyi	145
ornata }	Lophornis ornatus	82
ornatus }		
——	Orthorhynchus ornatus	117
orthura	Calliphlox amethystina	97
orthurus	Calliphlox amethystina	98
Osberti	Chlorolampis Osberti	174
Osryi	Phaëthornis Osryi	43
Otero	Leadbeatera Otero	71
Ourissia	Sporadinus? Maugei	173
pallidiceps	Heliomaster pallidiceps	131
Pamela }	Agloactis Pamela	107
Pamelæ }		
Pampa	Sphenoproctus curripennis	51
——	Sphenoproctus Pampa	51
paradisea	Panoplites flavescens	80
paradiseus	Topaza Pella	62
——, *Gmel.*	Not determined.	
parvirostris	Oxypogon Guerini	108
parvula	Agloactis parvula	108
Paradski	Heliotrypha Parzudaki	131
Parzudbaki	Sporadinus Riccordi	173
Paulinæ	Metallura tyrianthina	112
pavoninus	Phaëthornis, *Reichenb.*	
pectoralis	Lampornis gramineus	65
Pegasus	Chrysolampis moschitus	115
Pella	Topaza Pella	61
personatus	Schistes personatus	123
Peruana	Spathura Peruana	100
Peruanus	Chlorostilbon Peruanus	177

		PAGE
Petasophora	} Petasophora serrirostris	124
Petasophorus		
phænolæma		
phænolæma	Heliothrix phænolæma	121
phænoleuca		
Phaëthon	} Chlorostilbon Phaëthon	175
Phaeton		
Phaon	Cometes Phaon	104
Philippi	Phaëthornis Philippi	41
Phœbe, Less. et Delatt.	Not determined.	
Pichincha	Oreotrochilus Pichincha	63
pileatus	Orthorhynchus cristatus	117
pinicola	Heliomaster Leocadiæ	140
platura	} Discura longicauda	85
platurus		
platycerca	} Selasphorus platycercus	80
platycercus		
polytmus	Aïthurus polytmus	75
Poortmani	Panychlora Poortmanni	180
Popelairii	Prymnacantha Popelairei	80
porphyrogaster	Helianthea typica	130
porphyrura	} Lampornis porphyrurus	67
porphyrurus		
Ponchetti	} Heliothrix auriculatus	121
Ponchettii		
prasina	Chlorostilbon Atala	177
——	Chlorostilbon prasinus	176
prasinoptera	Eulampis jugularis	68
prasinus	Chlorostilbon chrysogaster	178
——	Chlorostilbon prasinus	176
——	Polytmus viridissimus	127
Pretrei	Phaëthornis superciliosus	45
Prevosti	Lampornis Prevosti	05
Primolii		
Primolina	} Metallura Primolii	112
Primolinus		
pristina	Amazilia pristina	155
Prunellei	} Lampropygia Prunellei	137
Prunelli		
puber	Chlorostilbon chrysogaster	178
Puchorani	Chlorostilbon prasinus	176
puella	Thalurania venusta	28
pulchra	Calothorax pulchra	91
punctatus	} Lampornis Mango	04
punctulatus		
puniceus	Orthorhynchus cristatus	117
purpuratus, Gmel.	Not determined.	
purpurea	Lampropygia purpurea	137
purpureiceps	Heliothrix Barroti	121
pygmæa	} Pygmornis pygmæa	49
pygmæus		
——	Pygmornis Aspasiæ	47
——	Mellisuga minima	87
——	Pygmornis ruficentris	48
Pyra	Topaza Pyra	62
quadricolor	Cyanomyia quadricolor	142
——	Lampornis Mango	05
Quitensis	Metallura Quitensis	112
radiosus	Cometes sparganurus	108
Raimondi	Sporadinus Ricordi	173
rectirostris	Doryfera rectirostris	71

		PAGE
recurvirostris	Avocettula recurvirostris	114
refulgens	Thalurania refulgens	77
Reginæ	Lophornis Reginæ	84
regia	Calliperidia Angelæ, *Reichenb.*	
Regulus	Lophornis Regulus	83
Reichenbachi	Chrysolampis moschitus	116
remigera	Spathura Underwoodi	100
Rhami	Lamprolæma Rhami	59
rhodotis	Petasophora iolata	125
Ricordi	Sporadinus Ricordi	173
Riefferi	Pyrrhophæna Riefferi	168
Rivoli	} Eugenes fulgens	58
Rivolii		
Roberti	Phæochroa Roberti	56
Robinson	Eustephanus Fernandensis	142
Rosæ	} Chætocercus Rosæ	92
Rosæ		
ruber	Selasphorus rufus	88
rubinea	} Clytolæma rubinea	131
rubineus		
rubinoides	Phæolæma Æquatorialis	143
——	Phæolæma rubinoides	142
rubra	Selasphorus rufus	88
Ruckeri	Glaucis Ruckeri	39
——	Glaucis Fraseri	39
rufa	Selasphorus rufus	88
ruficaudatus	Clytolæma rubinea	131
ruficaudus, *Vieill.*	Not determined.	
ruficeps	Ramphomicron ruficeps	100
ruficollis	Grypus nævius	35
rufigaster	Pygmornis Eremita	49
——	Pygmornis rufiventris	48
rufiventris	Pygmornis rufiventris	48
rufocaligata	} Spathura rufocaligata	100
rufocaligatus		
rufus	Campylopterus rufus	54
——	Selasphorus rufus	88
rutila	Pyrrhophæna cinnamomea	150
Sabina	} Adelomyia melanogenys	113
Sabinæ		
sagitta	Lesbiatura Otero	74
Salvini	Chlorolampis Salvini	174
sapphirina	Hylocharis lactea	171
——	Hylocharis sapphirina	171
sapphirinus	Hylocharis lactea	171
——	Hylocharis sapphirina	171
Sappho	Cometes sparganurus	103
Sasin	Selasphorus rufus	88
Saucerottei	Saucerottia typica	162
Saul		
Saulæ	} Lafresnaya Saulæ	70
Saulii		
scapulata	Eucephala scapulata	108
Schimperi	Circe latirostris	169
Schreibersii	Iolæma Schreibersi	73
scintilla	Selasphorus scintilla	89
Sclateri	Heliomaster Sclateri	132
scutatus	Augastes scutatus	123
sephanoides	Eustephanus galeritus	141
serrirostris	Petasophorus serrirostris	124
similis	Chlorostilbon Phaëthon	175

		PAGE
simpler	Aphantochroa cirrhochloris	55
——	Calothorax cyanopogon	91
——	Eriocnemis cupreiventris	144
Bitkensis	Selasphorus rufus	84
smaragdicaudus	Cynanthus Mocoa	103
smaragdina	Chlorostilbon chrysogaster	178
smaragdineum	Eucephala smaragdo-cærulea	169
smaragdinicollis	Metallura smaragdinicollis	113
smaragdinis	} Cynanthus Mocoa	103
smaragdinus		
smaragdo-cærulea	Eucephala smaragdo-cærulea	168
smaragdo-sapphiri- nus, *Shaw*	} Hylocharis cyanea, *Reichenb.*	
Sophiæ	Saucerottia Sophiæ	102
sordida	Phaoptila sordida	121
sparganura	} Cometes sparganurus	103
sparganurus		
spatuligera	Spathura Underwoodi	100
Spenceri	Heliangelus Spenceri	133
Spixi	Grypus Spixi	35
splendens	Campylopterus splendens	53
——	Campylopterus Villaviencio	53
——	Leadbeatera splendens	71
splendidus, *Vieill.*	Not determined.	
squalida	Phaëthornis squalidus	46
squalidus	Phaëthornis squalidus	45
squamata	Eriocnemis squamata	146
squamosa	Lepidolarynx meaolencus	140
squamosum	Grypus nævius	35
——	Lepidolarynx mesoleucus	140
Stanleyi	Ramphomicron Stanleyi	102
stellatus	Aithurus polytmus	75
stenura	Psnychlora stenura	180
Stokesi	Eustephanus Stokesi	142
striatus, *Gmel.*	Not determined.	
striigularis	Pygmornis striigularis	44
strophiana	} Heliangelus strophianus	132
strophianus		
strumaria	Lophornis magnificus	81
Stuartæ	Heliomaster Stuartæ	138
suavis	Pyrrhophæna Riefferi	123
Suecicus	Trochilus Alexandri	67
superba	Augastes scutatus	123
——	Heliomaster longirostris	138
superbus	Augastes scutatus	123
——	Heliomaster longirostris	138
superciliosus	Phaëthornis superciliosus	45
——	Phaëthornis malaris	41
——	Phaëthornis Preuri	45
——	Glaucis hirsuta	39
Surinamensis	Florisuga mellivora	80
——	Threnetes leucurus	40
——	Topaza Pella	62
Swainsoni	Sporadinus elegans	178
——	Doricha enicura	95
Sylphis	Lesbia Gouldi	111
syrmatophorus	Phaëthornis syrmatophorus	42
Temmincki	Lepidolarynx mesoleucus	140
——	Pterophanes Temmincki	105
Tondali	Calothorax, *Reichenb.*	
tephrocephala	Thaumatias albiventris	153

tephrocephalus	Thaumatias albiventris	153
thalassina	Petasophora thalassina	125
thalassinus	Petasophora Anais	124
——	Petasophora thalassina	125
Thalia, "Gould," Bsicà.	Unknown to me.	
Thaumantias	Polytmus viresceus	126
——	Thaumatias Linnæi	154
Thaumatias	Thaumatias albiventris	153
Thaurasia	Chrysobronchus viridicaudus	127
——	Polytmus viridissimus	127
Tobaci	}	
Tobagensis	Thaumatias Linnæi	153
Tobago	}	
Tomineo, Gmel.	Not determined.	
torquata	Bourcieria torquata	135
torquatus, Shaw	Not determined.	
tricolopha	Prymnacantha Popelairei.	
tricolor	Selasphorus platycercus	89
tristis	Patagona gigas	127
Tschudii	Thalurania Tschudii	78
Turneri	Leucippus chionogaster	150
typica	Lampropygia cœligena	136
——	Heliantbea typica	130
——	Saucerottia typica	162
——	Juliamyia typica	108
——	Myiabrillia typica	110
typicus	Saucerottia typica	162
typus	Phaethornis Guyi	44
tyrianthina	}	
tyrianthinus	Metallura tyrianthina	112
Underwoodi	Spathura Underwoodi	90
urochrysa	}	
urochrysia	Chalybura urochrysia	72
uropygialis	Eriocnemis vestita	145
varius, Gmel.	Not determined.	
ventilabrum	Spathura Underwoodi	90
venusta	Thalurania venusta	78
venustissimus	Eulampis jugularis	68
venustus	Augastes scutatus	123
Veraguensis	Lampornis Veraguensis	65
Verreauxi	Polemistria Verreauxi	85
versicolor	Cephalepis Delalandi	118
——	Thaumatias brevirostris	153
verticalis	Cyanomyia cyanocephala	147
——	Cyanomyia quadricolor	147
verticeps	Thalurania verticeps	78
vesper	}	
vespera	Rhodopis vespera	94
vestinigra	Eriocnemis nigrivestis	145
vestita	Eriocnemis cupreiventris	144
——	Eriocnemis vestita	145
vestitus	Eriocnemis vestita	145
Victoriæ	Lesbia Amaryllis	101
Vieilloti	Polemistria chalybea	85
——	Mellisuga minima	87
Vieillotii	Petasophora serrirostris	124
Villaviscensio	Campylopterus Villaviscencio	53
villosus	Phæthornis Oseryi	43
viola	Heliotrypha viola	131
violacea	}	
violaceus	Eulampis jugularis	68

			PAGE
violicauda	. . . .	Lampornis Mango	64
violiceps	. . . .	Cyanomyia violiceps	147
violifer	. . . .	} Helianthea violifera	131
violifera	. . . .		
violifrons	. . . .	Heliothrix violifrons	122
——	. . . .	Doryfera Johannæ	71
virescens	. . . .	Polytmus viridissimus	127
——	. . . .	Polytmus virescens	125
virginalis	. . . .	Lampornis virginalis	66
viridans	. . . .	Ailuruspolytmus	76
viridescens	. . . .	Polytmus virescens	125
viridicaudata	. . . .	Pygmornis Aspasia	47
viridicaudus	. . . .	Polytmus viridissimus	127
viridiceps	. . . .	Thaumatias viridiceps	152
viridigaster	. . . .	} Pyrrhophæna viridigaster	159
viridigastra	. . . .		
viridipallens	. . . .	Delattria viridipallens	60
viridipectus	. . . .	Thalurania nigrofasciata	78
——	. . . .	Thaumatias Linnæi	154
viridis	. . . .	Polytmus virescens	125
——	. . . .	Polytmus viridissimus	127
——	. . . .	Lampornis viridis	66
viridissima	. . . .	Thaumatias Linnæi	153
viridissimus	. . . .	} Polytmus viridissimus	127
——	. . . .	Thaumatias Linnæi	154
——	. . . .	Chlorostilbon prasinus	176
viridiventris	. . . .	Pyrrhophæna viridigaster	159
Vulcani	. . . .	Ramphomicron Vulcani	109
vulgaris	. . . .	Leucochloris albicollis	161
Wagleri	. . . .	Thalurania? Wagleri	79
Warszewiczii	. . . .	Diphlogena Aurora	134
——	. . . .	Saucerottia Warszewiczi	163
Watertoni	. . . .	Thalurania Watertoni	76
Wiedi	. . . .	Eucephala cyanogenys	107
Williami	. . . .	Metallura Williami	112
Wilsoni	. . . .	Lampropygia Wilsoni	137
Xantusi	. . . .	Heliopædica Xantusi	61
Yarrelli	. . . .	Myrtis Yarrelli	98
Yaruqui	. . . .	} Phaëthornis Yaruqui	44
Yaruqui	. . . .		
Yucatanensis	. . . .	Pyrrhophæna Yucatanensis	157
Zantusi	. . . .	Heliopædica Xantusi	61
Zonris	. . . .	Tryphæna Duponti	97
sonura	. . . .	Phæoptila sonura	170
——	. . . .	Pygmornis sonura	47

PROSPECTUS

OF THE WORKS

ON ORNITHOLOGY,

AND ON

THE MAMMALIA OF AUSTRALIA,

BY

JOHN GOULD, F.R.S., ETC.

All the Author's works are in Imperial Folio, forming a regular series. They are—

I. A CENTURY OF BIRDS FROM THE HIMALAYA MOUNTAINS. 1 Vol. Imperial Folio, containing 80 Plates, with descriptive letter-press. Price £14 14s. London 1832.

 This work, of which no copies remain, was commenced in January 1831, and completed in August 1832. It contains figures and descriptions of 100 Birds on 80 Plates which were at that time either new or very imperfectly known.

II. THE BIRDS OF EUROPE. 5 Vols. Imperial Folio, comprising 449 Plates, with descriptive letter-press, Introduction, &c. Price £78 6s. London 1837.

 The whole of the copies of this work have been disposed of; and when any one of them is offered for sale on the demise of a Subscriber, or from other causes, it realizes considerably more than its original cost. Thus in the year 1860 a bound copy was sold for £120, and an unbound one realised £104.

III. A MONOGRAPH OF THE RAMPHASTIDÆ, OR FAMILY OF TOUCANS. 1 Vol. Imperial Folio, containing Fifty-two Plates, with descriptive letter-press, &c. Price £12 12s. London 1854.

 An edition of this work was published in 1834 at the price of £7; but the extensive researches carried on during the last twenty years among the Great Andean Ranges of South America having led to the discovery of many additional and beautiful species belonging to this extraordinary group of Birds, a revision of the work not only became necessary, but an entirely new edition was deemed imperative. This edition, comprising every known species, with the whole of the Plates redrawn, and with an Introduction containing much valuable information derived from authentic sources, was published in 1854 at the price of £12 12s.

 The history of this South American group is very peculiar; and their manners and actions are as remarkable as their aspect; in some respects reminding us of the Hornbills of India and Africa, while in others they are unlike those of any other group.

IV. A MONOGRAPH OF THE TROGONIDÆ, OR FAMILY OF TROGONS. 1 Vol. Imperial Folio, containing Thirty-six Plates, with descriptive letter-press. Price £8. London 1838.

This work, like the Monograph of the Toucans, comprises the history and figures of all the species of the group known up to the date of publication. The members of the Trogonidæ are remarkable for a gorgeous style of colouring, for their recluse habits, and for the union of insect diet with such aliments as fruits and berries, in accordance with which the beak is modified; they are divided between the warmer latitudes of America, India, and the adjacent islands, with the exception of one species, which is peculiar to Africa.

The same reasons which induced the Author to publish a new edition of the Monograph of the Ramphastidæ, have also rendered another edition of this Monograph desirable; and accordingly one is now in preparation, comprising all the new species and information acquired respecting this family of birds during the last twenty years. It will be completed in four Parts, price £3 3s. each, the first of which is now ready for delivery.

V. THE BIRDS OF AUSTRALIA. 7 Volumes, Imperial Folio, containing Figures of 600 species, with descriptive letter-press and a large amount of Introductory matter. Price £115. London 1848.

This work was originally published in Thirty-six Parts, each containing Seventeen Plates with descriptive letter-press, at the price of Three Guineas each Part, with the exception of the Thirty-sixth, the price of which, in consequence of the large amount of introductory matter, is £4 12s.

The Birds of Australia, comprised in seven handsome folio volumes, is considered by the Author as the most important and original work which he has yet published. It contains the Ornithology of one vast portion of the globe, and that portion of no small importance in whatever point of view it be considered. Impressed with the necessity of rendering this tribute to science worthy of acceptation, the Author left England for Australia in May 1838; and after remaining there for two years in order to study the habits and manners of the birds and quadrupeds, he returned with a great amount of novel, strange, and interesting facts. The habits of the Bower-birds, the Mound-makers (*Talegalla*, *Leipoa*, &c.), and of the Lyre-bird, when made known, were deemed especially marvellous; but every statement has been subsequently confirmed to the letter. To dilate upon the peculiarities of the Fauna and Flora of Australia is not the Author's present aim. Suffice it to say, he has endeavoured to the utmost to do justice to this work to its Ornithology; and so well have his labours been received, that very few copies of this great work remain on his hands for disposal, and ere long, like the "Birds of Europe," they will be at a premium.

As the at present unexplored portions of Australia become more and more known, additional species of birds will doubtless be discovered, rendering a Supplement to the work necessary, in order to keep the subject complete; and this will be issued in Parts as a sufficient number of novelties come to hand: thus, a portion of the new and interesting species lately brought home by the naturalist and officers of several of H.M. surveying ships, and some derived from other sources, have appeared under the title of "Birds of Australia," Supplement, Parts I., II., III., price £3 3s. each; and any other novelties that may arrive will in like manner be published, and when a sufficient number of parts to form a volume have been issued, a Title-page and every other requisite will be supplied.

VI. A MONOGRAPH OF THE ODONTOPHORINÆ, OR PARTRIDGES OF AMERICA. 1 Vol. Imperial Folio, containing Thirty-two Plates, with descriptive letter-press. Price £9 8s. London 1850.

The interest which attaches to this work is threefold. First, it displays, even to the most unpractised eye, the broad distinction which subsists between the Partridges of America and those of Europe; secondly, the species are all remarkable for the elegance of their forms and for the chaste beauty of their colouring; and thirdly, at no distant date these Birds will doubtless be regarded in America as our Partridges in Europe are, as game, and perhaps preserved by law—their flesh being as delicate for the table as that of our ordinary bird, from which, however, they differ considerably in the structure of the beak, and in their habits and economy.

VII. THE BIRDS OF ASIA.

To no portion of the globe does there attach so much interest as to that vast extent of the Old World which we designate Asia. It is there that all the productions of nature essential to the well-being of man occur in the greatest abundance. The most important of our domestic quadrupeds, the most valuable and interesting of our domestic Gallinaceous birds, were first reclaimed in Asia. That the Zoology, then, of such a country should have called forth the notice and study of able minds cannot be surprising; and yet it is remarkable that no one has attempted a work comprehending a general history of its ORNITHOLOGY. This hiatus in Ornithological literature the Author proposes to fill up by publishing a work on "The Birds of Asia," precisely similar in every respect to his former works on "The Birds of Europe" and "The Birds of Australia." Its size and manner of execution will be the same; and it will be published in Parts, price Three Guineas each.

Of this work thirteen Parts are published; and for the present it will appear at the rate of not more than one or two Parts a year.

VIII. A MONOGRAPH OF THE TROCHILIDÆ, OR HUMMING-BIRDS.

Having from an early period devoted himself to the study of these beautiful birds, and having acquired a most valuable and extensive collection of a group essentially peculiar to America and its adjacent islands, the Author determined upon publishing a Monograph of a family unequalled for the gorgeous and ever-changing brilliancy of their hues, the variety of their form, the singularity of their habits, and the extent of their territorial distribution. Anxious to render his representations of these lovely objects as faithful as possible, the Author instituted a series of experiments upon a new mode of colouring, which has been so far successful, that the birds are as closely imitated as art can hope to see accomplished; he has also endeavoured, as far as possible, to associate them with the plants of its own region, thereby adding an additional charm to a work which he trusts will be equally acceptable to the artist and the lover of nature, and which has been so successful that it bids fair to be the most popular of his productions.

This Monograph is now complete in 25 Parts, forming five volumes, in which 300 species are figured: 24 of these parts contain Fifteen Plates each, with descriptive letter-press, and the 25th Title-pages, Introduction, &c. The price of each part is £3 3s. The copies remaining unsubscribed for may be had complete, or for the convenience of future subscribers at the rate of 5 Parts a year, commencing with January 1862, in which case they will be required to complete their copies, as an equal number of all the Parts have been printed, and the drawings effaced from the stones; this work, therefore, like its predecessors, will shortly become scarce.

IX. ICONES AVIUM, or Figures and Descriptions of new and interesting Species of Birds from various Parts of the World, forming a Supplement to the Author's other Works.

The object of this Work is explained in the Title: It will be issued as novelties of interest occur, in Imperial Folio Parts containing ten species with descriptive letter-press, price £1 15s. each. Two Parts have been published, one in 1837, the other in 1838.

X. THE MAMMALS OF AUSTRALIA.

The Author's visit to Australia having enabled him to procure much valuable information respecting the habits and economy, and many new species, of the singular and interesting Mammalia of that country, he has determined upon publishing a work on the subject. With respect to the importance of such a work no doubt can exist; and as the Author is deeply impressed with this idea, so will he endeavour to render it equal to its associate publication on the Ornithology of that remarkable region.

In its execution this work will be precisely similar to the "Birds," and will be completed in Thirteen or Fourteen Parts, each containing Fifteen Plates, price £3 3s. each.

Twelve parts have been published, and have been so highly approved of, that by many they are regarded as more interesting than the "Birds." The Thirteenth Part is now in preparation, and the work is consequently approaching its close.

The Author begs to add that, when possible, he will be happy to perfect any sets at present incomplete, upon the possessors communicating to him at the following address their wish on the subject.

LONDON: PUBLISHED BY THE AUTHOR AT 20 CHARLOTTE STREET, BEDFORD SQUARE, W.C.

September 1, 1861.

www.ingramcontent.com/pod-product-compliance
Lightning Source LLC
Chambersburg PA
CBHW021304240426
43669CB00042B/1196